Total Quality Management
An Introductory Text

Paul T J James

Prentice Hall
London   New York   Toronto   Sydney   Tokyo   Singapore
Madrid   Mexico City   Munich

First published 1996 by
Prentice Hall Europe
Campus 400, Maylands Avenue
Hemel Hempstead
Hertfordshire, HP2 7EZ
A division of
Simon & Schuster International Group

© Paul T.J. James 1996

All rights reserved. No part of this publication may be reproduced,
stored in a retrieval system, or transmitted, in any form, or by any
means, electronic, mechanical, photocopying, recording or otherwise,
without prior permission, in writing, from the publisher.
For permission within the United States of America
contact Prentice Hall Inc., Englewood Cliffs, NJ 07632

Typeset in 10/12 pt Garamond Book
by MHL Typesetting Ltd

Printed and bound in Great Britain by
T.J. Press (Padstow) Ltd

Library of Congress Cataloging-in-Publication Data

James, Paul T. J.
　　Total quality management: an introductory text/Paul T.J. James.
　　　p.　cm.
　　Includes bibliographical references (p.　) and indexes.
　　ISBN 0-13-207119-3
　　1. Total quality management.　I. Title.
HD62.15.J35　1996　　　　　　　　　　　　　　　　　　95-39465
658.5′62—dc20　　　　　　　　　　　　　　　　　　　　CIP

British Library Cataloguing in Publication Data

A catalogue record for this book is available from the British Library

ISBN 0-13-207119-3

1　2　3　4　5　　00　99　98　97　96

To BOD.

BRIEF CONTENTS

Part 1	Quality Management — The Basics	7
	Chapter 1 The basics of management	9
Part 2	Quality Management Concepts	37
	Chapter 2 Total quality management	39
	Chapter 3 Quality management writers	61
	Chapter 4 The three views of quality	79
Part 3	The Five Functions of Quality Management	91
	Chapter 5 Quality planning	93
	Chapter 6 Quality of design	113
	Chapter 7 Organizational structure and design	132
	Chapter 8 Leadership	142
	Chapter 9 Group dynamics	165
	Chapter 10 Human resource management	188
	Chapter 11 Culture and change management	203
	Chapter 12 Control	229
	Chapter 13 Statistical process control	245
	Chapter 14 Quality economics	275
	Chapter 15 Quality standards	283
Part 4	Integrated Quality Management — The Future	315
	Chapter 16 Total quality management: Future issues that need to be addressed today	317
Part 5	Cases and Problems	327
	INFOCOM International: Customer service	329
	ALPHA-ONE: Changing the quality-oriented culture of an organization	332
	Lotteries Commission: Managing quality with data	338
	Quality improvement tools and BOD Inc.: A football team	341

Problems	343
Recent research	349
Quality glossary	354
Bibliography	362
Name index	369
Subject index	372

CONTENTS

ACKNOWLEDGEMENTS		xxi
INTRODUCTION		1
PART 1	QUALITY MANAGEMENT — THE BASICS	7
Chapter 1	**The basics of management**	9
	CHAPTER OBJECTIVES	9
	CHAPTER OUTLINE	9
	1.1 Introduction	10
	1.2 Definitions	10
	1.3 Levels of management	11
	1.4 What managers do	11
	1.5 Managerial skills	13
	1.6 Managerial roles	14
	1.6.1 *Interpersonal roles*	15
	1.6.2 *Informational roles*	16
	1.6.3 *Decision-making roles*	16
	1.7 The development of management theory	17
	1.8 The theories of management	18
	1.8.1 *The classical theories of management*	19
	The functions of management	24
	Level of the use of the functions of management	25
	The management process	26
	1.8.2 *The behavioural theories of management*	27
	The pre-behavioural school theorists	28
	The behavioural school	29
	1.8.3 *The human relations theories of management*	31
	1.8.4 *The systems theories of management*	32
	1.8.5 *The contingency theories of management*	33

	1.9	Chapter review	34
	1.10	Chapter questions	35

PART 2 — QUALITY MANAGEMENT CONCEPTS — 37

Chapter 2 — Total quality management — 39

CHAPTER OBJECTIVES — 39
CHAPTER OUTLINE — 39

2.1	Introduction	40
2.2	The four eras of quality management	40
	2.2.1 *Quality development through inspection*	40
	2.2.2 *Quality development through quality control*	43
	2.2.3 *Quality development through quality assurance*	44
	2.2.4 *Quality development through TQM*	45
	Implementation issues	48
	Management functions and the issues to be considered when implementing TQM	51
	The process of the implementation of TQM	52
	TQM practices — the reality	55
2.3	Chapter review	58
2.4	Chapter questions	60

Chapter 3 — Quality management writers — 61

CHAPTER OBJECTIVES — 61
CHAPTER OUTLINE — 61

3.1	Quality management writers	61
	3.1.1 *Juran*	62
	3.1.2 *Deming*	64
	3.1.3 *Garvin*	70
	3.1.4 *Crosby*	70
	3.1.5 *Ishikawa*	73
	3.1.6 *Feigenbaum*	75
	3.1.7 *Taguchi*	75
3.2	Chapter review	76
3.3	Chapter questions	77

Chapter 4 — The three views of quality — 79

CHAPTER OBJECTIVES — 79
CHAPTER OUTLINE — 79

	4.1	Introduction	80
	4.2	The three quality views	80
	4.3	The five quality bases	81
		4.3.1 *Transcendent quality view*	81
		4.3.2 *Product-based quality view*	81
		4.3.3 *User-based quality view*	82
		4.3.4 *Manufacturing-based quality view*	82
		4.3.5 *Value-based quality view*	82
	4.4	Where does the customer fit into all this?	83
	4.5	The effect of the differing views of quality	83
	4.6	Factors affecting customer perceptions of quality	84
		4.6.1 *Performance*	85
		4.6.2 *Features*	85
		4.6.3 *Reliability*	85
		4.6.4 *Conformance*	86
		4.6.5 *Durability*	87
		4.6.6 *Serviceability*	87
		4.6.7 *Aesthetics*	87
		4.6.8 *Perceived quality*	88
	4.7	A comparison of the five quality bases and the eight dimensions	88
	4.8	Chapter review	89
	4.9	Chapter questions	89
PART 3		**THE FIVE FUNCTIONS OF QUALITY MANAGEMENT**	**91**
Chapter 5		**Quality planning**	**93**
		CHAPTER OBJECTIVES	93
		CHAPTER OUTLINE	93
	5.1	Introduction	94
	5.2	Why plan?	94
	5.3	Why the need for planning for quality?	95
	5.4	Who is responsible for planning for quality?	95
	5.5	Types of quality plans	96
	5.6	What is quality planning?	96
	5.7	The quality planning process	97
		5.7.1 *Environmental analysis*	98
		5.7.2 *Quality mission*	98
		5.7.3 *Setting a quality policy*	99
		5.7.4 *Generate strategic quality goals*	100
		5.7.5 *Establish quality action plans*	102
		5.7.6 *Quality strategy implementation*	102

		5.7.7	Monitor and evaluate quality performance	103
	5.8	Benchmarking		104
		5.8.1	Definitions of benchmarking	104
		5.8.2	Goal setting and benchmarking	105
		5.8.3	Types of benchmarking	106
		5.8.4	Characteristics or indicators used in developing benchmarking practices	106
		5.8.5	Influences on the benchmarking process	107
		5.8.6	The benchmarking process	107
		5.8.7	Benefits of benchmarking	108
		5.8.8	Benchmarking limitations	109
	5.9	Chapter review		110
	5.10	Chapter questions		112

Chapter 6 **Quality of design** 113
CHAPTER OBJECTIVES 113
CHAPTER OUTLINE 113

	6.1	Introduction		114
	6.2	Marketing and design		114
	6.3	Who is the customer?		115
	6.4	The quality service culture — providing customer service		117
		6.4.1	Quality service effectiveness	118
		6.4.2	Managing the quality of service offered	119
		6.4.3	Benefits of the application of quality of service	120
	6.5	Marketing planning		121
		6.5.1	The marketing mix	121
	6.6	Some tools and methods used to ensure conformance to customers' needs and wants		122
	6.7	Benefits of customer-oriented marketing by ensuring fitness for purpose of goods supplied		123
	6.8	Traceability		123
		6.8.1	Control documentation used to effect traceability	124
		6.8.2	The effect of standards on traceability	125
		6.8.3	Some of the perceived problems of traceability	126
	6.9	Quality function deployment		126
		6.9.1	QFD mechanics	128
	6.10	Chapter review		129
	6.11	Chapter questions		131

Chapter 7	Organizational structure and design	132
	CHAPTER OBJECTIVES	132
	CHAPTER OUTLINE	132
	7.1 Organizing	132
	7.2 Organizational structure	133
	7.3 Organizational design	134
	7.4 Job design	136
	7.4.1 *Job design methods*	136
	7.5 Centralization versus decentralization	139
	7.6 Implications for organizational effectiveness	140
	7.7 Chapter review	140
	7.8 Chapter questions	141
Chapter 8	**Leadership**	**142**
	CHAPTER OBJECTIVES	142
	CHAPTER OUTLINE	142
	8.1 What is quality leadership?	143
	8.2 Strategic quality leadership — the need	144
	8.3 How leaders influence — the power stakes	144
	8.4 Theories of leadership	145
	8.4.1 *The trait qualities of leadership*	145
	8.4.2 *The behavioural qualities of leadership*	146
	8.4.3 *The situational qualities of leadership*	148
	8.4.4 *Self-leadership*	152
	8.5 Motivation	154
	8.5.1 *The nature of motivation*	154
	8.5.2 *Theories of motivation*	155
	The needs (content) theories	155
	The cognitive (process) theories	160
	The reinforcement theories	162
	8.6 Chapter review	163
	8.7 Chapter questions	164
Chapter 9	**Group dynamics**	**165**
	CHAPTER OBJECTIVES	165
	CHAPTER OUTLINE	165
	9.1 Introduction	166
	9.2 What is a group?	166
	9.3 Characteristics of a group	166
	9.4 Types of groups found in organizations	167
	9.4.1 *Formal groups*	167
	9.4.2 *Informal groups*	168
	9.5 Group development	169
	9.6 Group effectiveness and efficiency	172

	9.6.1	*Group inputs*	172
		Group composition	172
		Group appeal	172
		Group roles	173
		Group size	173
	9.6.2	*Group process*	173
		Group norms and conformity	174
9.7	Team building		176
9.8	Conflict		177
	9.8.1	*Managing conflict*	179
9.9	Communication		180
	9.9.1	*Why communication is important in groups*	180
	9.9.2	*Types of communication*	182
		Verbal communication	182
		Non-verbal communication	182
	9.9.3	*The communication process*	183
	9.9.4	*Group communication configurations*	184
		Centralized configurations	184
		Decentralized configurations	185
9.10	Chapter review		185
9.11	Chapter questions		187

Chapter 10 **Human resource management** **188**
CHAPTER OBJECTIVES 188
CHAPTER OUTLINE 188
10.1 Introduction 189
10.2 Definition of human resource management (HRM) 189
10.3 HRM and TQM 189
10.4 HRM planning 191
10.5 Recruitment 192
10.6 Selection 193
10.7 Training, education and development of employees 195
10.8 Performance appraisal 197
10.9 Compensation 197
10.10 Workforce relationships and HRM 198
10.11 Quality circles 199
10.12 Chapter review 200
10.13 Chapter questions 202

Chapter 11 **Culture and change management** **203**
CHAPTER OBJECTIVES 203
CHAPTER OUTLINE 203

	11.1	The nature of change	204
		11.1.1 *Introduction*	204
		11.1.2 *Definitions of change*	204
		11.1.3 *Causes of change in organizations*	205
		11.1.4 *What is the best way to deal with change?*	206
		11.1.5 *What is the best way to implement change?*	208
		11.1.6 *Resistances to change programmes*	209
		11.1.7 *Change implementation process*	212
		11.1.8 *The use of the change agent*	214
		Change agent styles	215
		11.1.9 *Process consultation*	216
		11.1.10 *Change interventions*	218
		11.1.11 *Change and the organizational lifecycle*	219
	11.2	Culture — group and organizational	219
		11.2.1 *Definition of culture*	219
		11.2.2 *Change and the need and use of power in organizations*	221
		11.2.3 *The politics of work relationships*	223
	11.3	Chapter review	226
	11.4	Chapter questions	228
Chapter 12	**Control**		**229**
	CHAPTER OBJECTIVES		229
	CHAPTER OUTLINE		229
	12.1	What is control?	230
	12.2	Quality control systems	230
	12.3	Control process requirements	232
		12.3.1 *Choosing what to control — the subject*	232
		Performance indicators and their relationship to quality control subjects	233
		12.3.2 *Developing a target for a control characteristic*	233
		12.3.3 *Determining a unit of measure*	234
		12.3.4 *Developing a means to measure the control characteristic*	234
		12.3.5 *Measuring the characteristic in the production arena or field*	235
		12.3.6 *Evaluating the difference between actual and expected performance*	235
		12.3.7 *Taking action as necessary*	236

xvi CONTENTS

12.4	Materials control and prevention	236
12.4.1	*JIT methods*	236
12.5	The seven old and seven new tools of quality management	237
12.5.1	*The seven old quality tools*	237
	Flowcharts	237
	Check sheets	238
	Histograms	239
	Cause and effect diagrams (fishbone diagrams)	239
	Pareto diagrams	240
	Scatter diagrams	242
	Control charts	242
12.5.2	*The seven new quality tools*	242
	Affinity diagrams	242
	Interrelationship diagraphs	243
	Tree diagrams	243
	Matrix diagrams	243
	Matrix data analysis	243
	Arrow diagrams	243
	Process decision programme chart	243
12.6	Chapter review	243
12.7	Chapter questions	244

Chapter 13 Statistical process control 245

CHAPTER OBJECTIVES 245
CHAPTER OUTLINE 245

13.1	Introduction	246
13.2	What is statistical process control (SPC)?	246
13.3	The process of inspection	247
13.3.1	*Measurement*	247
13.3.2	*Quality measurements*	248
13.3.3	*Variation*	248
13.4	Statistical control charts	249
13.4.1	*The concept of the control chart (Shewhart control chart)*	249
	Definitions	249
	Application of the control chart	250
13.4.2	*Steps in the development of a control chart*	252
13.5	Quality control charts	253
13.5.1	*Variable control charts*	253
13.5.2	*Attribute control charts*	256
13.6	Cumulative summation control chart (CuSum)	260
13.7	Comparison of the control and CuSum charts	263

	13.8 Advantages and disadvantages of control charts	263
	13.9 Advantages and disadvantages of CuSum charts	263
	13.10 Acceptance sampling	264
	13.11 Acceptance plans	265
	13.12 Sampling plans	267
	13.12.1 *The single sampling plan*	267
	13.12.2 *The double sampling plan*	268
	13.13 The operating characteristic curve	268
	13.13.1 *Constructing the operating characteristic curve*	269
	13.14 Process capability	269
	13.14.1 *Process capability uses*	269
	13.15 SPC and quality improvement — what it means	270
	13.16 Process variability reduction	271
	13.17 Taguchi — the quality loss function	272
	13.18 Chapter review	273
	13.19 Chapter questions	274
Chapter 14	**Quality economics**	**275**
	CHAPTER OBJECTIVES	275
	CHAPTER OUTLINE	275
	14.1 Introduction	275
	14.2 What are quality-related costs?	276
	14.3 Classification of quality costs	276
	14.4 The importance of quality costs to the quality-oriented organization	277
	14.5 Quality costs — why measure them?	278
	14.6 Cost of quality versus cost of non-quality	279
	14.7 Hidden costs of quality	279
	14.8 Lifecycle costs	280
	14.9 The management of quality costs	280
	14.10 Chapter review	282
	14.11 Chapter questions	282
Chapter 15	**Quality standards**	**283**
	CHAPTER OBJECTIVES	283
	CHAPTER OUTLINE	283
	15.1 Introduction	284
	15.2 What is a quality system?	284
	15.3 BS EN ISO 9000	284
	15.3.1 *Origin and basis*	285
	15.3.2 *The parts of BS EN ISO 9000*	287
	15.3.3 *BS EN ISO 9000 — update 1994*	289
	General	289
	Specific changes and implications	290

15.4 Certification and accreditation to
BS EN ISO 9000 291
    15.4.1 *Definitions* 292
    15.4.2 *Certification* 292
        First party assessment 292
        Second party assessment 293
        Third party assessment 294
    15.4.3 *Benefits of certification to BS EN ISO 9000* 294
        General audit process for certification 294
    15.4.4 *Accreditation* 295
        NACCB — National Accreditation Council for Certification Bodies 296
        Certification — scope 296
        Accreditation — scope of a certificating body 297
        The process of accreditation 298
15.5 Auditing to BS EN ISO 9000 299
15.6 BS 7850 — total quality management 301
15.7 Other standards that may be useful to consult 302
    15.7.1 *BS 7750 — environmental management standard (ISO 1996)* 302
    15.7.2 *Investors in People (IIP)* 303
    15.7.3 *European Quality Award (EQA)* 304
15.8 Implications of the application of BS EN ISO 9000 — a service approach — higher education 304
15.9 Chapter review 307
15.10 Chapter questions 309
Appendix 1 — BS EN ISO 9000 Quality System Elements 310
Appendix 2 — BS EN ISO 9000 Survey Instrument 311
Appendix 3 — EN 45012 Clauses 313

PART 4   INTEGRATED QUALITY MANAGEMENT — THE FUTURE   315

Chapter 16   Total quality management: Future issues that need to be addressed today   317

CHAPTER OUTLINE 317
16.1 Introduction 317
16.2 Integrated systems of quality management 318
    16.2.1 *Processes* 318
    16.2.2 *People* 320
    16.2.3 *Structures* 322

	16.2.4 *Technology*	322
	16.2.5 *Customers*	323
	16.3 Ideological basis for future effective development and operation of TQM	324
	16.4 Concluding remarks	325

PART 5   CASES AND PROBLEMS   327

INFOCOM International: Customer service   329
Activity brief   331

ALPHA-ONE: Changing the quality-oriented culture of an organization   332
Introduction   332
Background   332
The Situation   333
     The change process   333
     Culture change   334
     Impact of the change programme   335
     1993 and beyond — What activity occurred?   335
         Strategies   336
Conclusion   337
Activity brief   337

Lotteries Commission: Managing quality with data   338
Introduction   338
Activity brief   340

Quality improvement tools and BOD Inc.: A football team   341
Background   341
Situation   341
The brainstorming session results   342
The next step   342
Activity brief   342

Problems   343
Problem No. 1   343
Problem No. 2   343
Problem No. 3   343
Problem No. 4   343
Problem No. 5   344
Problem No. 6   344
Problem No. 7   344

Problem No. 8	344
Problem No. 9	344
Problem No. 10	344
Problem No. 11	344
Problem No. 12	345
Problem No. 13	345
Problem No. 14	346
Problem No. 15	346
Problem No. 16	346
Problem No. 17	346
Problem No. 18	347
Recent research	**349**
The results of a small business BS EN ISO 9000 survey in South Wales	349
Concerns raised in the piloting of the survey	349
Problems of the small business inquiry	349
Methodology	350
Procedure	350
Response rate of survey	350
Findings	351
Statement of outcomes of the ISO 9000 research	353
Conclusion	353
Quality glossary	**354**
Bibliography	**362**
Name index	**369**
Subject index	**372**

ACKNOWLEDGEMENTS

It takes many individuals to provide the professionalism and support in order to develop and produce a product that satisfies the final consumer. The development and preparation of this text is by no means different.

I am indebted to the many contributors that have helped to make the production of this text possible. Thanks go to my colleagues for their professionalism, to my students for their questioning and enthusiasm, and to the business community that provided the backdrop for the quality focus.

Thanks must also go to the publishing team at Prentice Hall who have encouraged me through the publishing quagmire. Specifically, to John Yates my sponsoring editor, who has worked incessantly through the evolution of the project. Thanks must also be given to my reviewers.

Every effort has been made to contact copyright holders. When the appropriate information has been received we shall be pleased to make the necessary acknowledgement at the first opportunity.

INTRODUCTION

Customer support, customer orientation and customer satisfaction are the major objectives for the quality-oriented organization. Total quality management (TQM) is essentially about the development of an ideology, a philosophy, methods and actions that are designed to satisfy customers completely, through their continuous improvement. Managers are attempting to apply the above philosophies and principles to do just this, and the results can be quite astounding.

However, TQM is more than this; it is a way of life — a way of a working life. Because of increasing competition in most, if not all, commercial sectors, attention to quality in order to satisfy the requirements of consumers' wants and needs signifies a major change in the way organizations are managed. The quality-oriented organization is outward oriented, flexible in approach and practice, nurtures its staff, suppliers and customers and provides more than just a job.

The challenge for top management is to determine how and when to let staff run the part of the business that they know most about. The application of quality management allows staff to become their own managers, to lead themselves, and to become experts in carrying out their tasks.

Quality management therefore has a great burden placed on it. The first is to satisfy customers continuously; the second is to provide a learning environment within the organization; and the third is to ensure the survival of the organization. These can all be accomplished through using TQM and this challenge will not go away as competitors take this challenge into the new millennium.

The text concept

The text provides the basis for a quality management textbook. It is a general text, whose main purpose is to provide a measure of depth to any related quality issue within an organizational setting. The orientation of the text is towards both manufacturing and services *per se*, and tries to provide a balanced perspective in relation to quality issues in both. This text provides an integrated approach to the application of the five functions of TQM — planning, organizing, leading, controlling and staffing — and develops a theoretical and practical basis to effect this.

The reader of this text is considered to be a student of quality management, whether student, researcher, professor, manager or someone possibly interested in understanding the context, depth, scope and implications of TQM today.

Quality management is being taught in business schools, colleges, universities, other places of learning and in commercial or industrial organizations. What is generally taught are programmes reflecting the mechanics of quality control or quality assurance. The scope of this text extends this into areas considered absolutely necessary, such as quality management systems, management of change, the importance, implications and practice of marketing and design, benchmarking and customer service. Today's quality practitioner cannot be an isolated professional and this text therefore provides knowledge of many of the related management practices that are deemed a minimum requirement.

Much of the basis for today's quality management practices can be seen in general management theory. In this context, much research time has been spent on re-evaluating original thoughts, ideas and notions developed in earlier management texts. The result is the grounding of quality management theory from many of the original writers, rather than from later sources. In this respect, the authenticity of the text treatment is captured.

This book is designed to be suitable as a main text in a postgraduate course — MBA, DMS, etc. — or an undergraduate management or business course; or indeed any course of study that has a module specifically involving quality management. It is intended that the book will provide a wider approach to quality and integrate general management and quality management theories in order to provide a cohesive treatise. Significantly, the text takes individuals from the basic management theories through to modern-day quality management and beyond.

It is therefore a text that allows students or business personnel with little or no knowledge of quality to grasp quality theories effectively, understand and apply these theories in practice. It also provides good background reading and consolidation for quality professionals and practitioners.

Pedagogy and learning features

- *Objectives and learning outcomes for each chapter* — these are designed to indicate the necessary learning outcomes that a student can expect to gain from effectively working through the chapter content.
- *Chapter outline at beginning of chapter* — this suggests the main themes or general areas that may be encountered within the chapter.
- *End of chapter review* — this provides a potted version of the chapter content.
- *End of chapter questions* — these are designed to provide a means in which to focus study of the varied mix of chapter material.
- *Cases and problems section at the rear of the text* — this is included to ensure

a more integrated concentration of the problems and issues developed out of the text topics.
- *Quality glossary* — this is included in order to provide a means quickly to evaluate the developed meanings and concepts in the text.
- *Name and subject indexes at rear of text* — these are included as part of a traditional approach for name and subject reference.

Text organization

This text can be characterized as traditional in layout but contemporary in content and thus parallels the development of general management theory with TQM theory.

Therefore, much of the basis for TQM in this text is formulated to take advantage of these related trends, and is an ongoing, continuous development in the application of management to organizations.

This text is divided into three parts. Part 1 — Chapter 1 — deals with the basics of general management theory. Part 2 — Chapters 2, 3 and 4 — discusses the concepts of TQM, the foundation writers on quality and the implications of the various developed views on quality. Part 3 — the remaining chapters of the text — discusses the five functions of TQM — planning, organizing, leading, staffing and controlling quality in organizations. The text ends in the brief presentation of the contextual issues that confront the future development of an integrated TQM system.

The structural layout of the chapters provides an integrated and extending basis for the application of TQM. In this regard, theoretical concepts are dealt with in the initial chapters, and more practical issues and techniques later in the text. This is a deliberate ploy to offer the theoretical context and understanding before using practical quality management methods.

Overview of the chapters

Chapter 1 focuses on the *basics of management*. Here, basic management terminology, the development of management theory, the various schools of management, the functions of management and the management process are discussed.

Chapter 2 discusses the main issues facing the development and use of *total quality management*. An examination of the facets of TQM is carried out, which answers the question, What is TQM?, and the implications for the implementation of TQM to manufacturing and services are explored.

Chapter 3 focuses on the *quality management writers* and evaluates their contributions throughout this century — including those from the United Kingdom, United States and Japan.

Chapter 4 evaluates the impact of the *three views of quality*. In particular, it discusses the effects of the various quality views and compares them to the eight dimensions of quality.

Chapter 5 provides the basis for *developing a quality mission, goals and objectives*. The chapter appraises the application and need of quality planning practices in organizations, evaluates the quality planning process, and determines the basis and use of benchmarking as a quality planning tool.

Chapter 6 covers the major *quality of design* issues facing organizations and discusses the related and necessary integration of the disciplines of marketing and design and how these can be used to ensure customer satisfaction.

Chapter 7 evaluates the effect of *organizational structure and design* on quality-related issues. Job design methods are discussed and comparisons of the centralization versus decentralization structures are made.

Chapter 8 focuses on *leadership* as an important topic in quality management. Here, how leaders influence is discussed, as are the various theories of leadership, and how these affect the application of quality-related activities.

Chapter 9 develops the major concepts related to *group dynamics*. A description of the characteristics and types of groups is made, an evaluation of group efficiency and effectiveness, and the needs and techniques of conflict management are discussed.

Chapter 10 highlights the increasing respect given to *human resource management (HRM)* in quality-oriented organizations. The relationship between HRM and TQM is discussed, as is the HRM planning process, recruitment and selection, performance appraisal, compensation issues, and the impact of quality management on workforce relationships.

Chapter 11 evaluates the development and implications of *culture and change and management*. It discusses change and its meaning in a quality-oriented organization, cultural implications of managing change, and the need to appraise the impact of the various issues surrounding the use of power in organizations.

Chapter 12 develops the theoretical concepts of *control*. It evaluates control systems, control requirements, performance indicators and their relationship to managing quality, and discusses the seven old and new tools.

Chapter 13 examines the use, techniques and requirements of *statistical process control*. Quality control charts methods are discussed, as are sampling, process capability techniques, and the relationship between statistical process control (SPC) and quality.

Chapter 14 highlights the need to address *quality economics* and determines the basis and importance for understanding and evaluating quality-related costs.

Chapter 15 evaluates the implications for implementing *quality standards* — management and environmental. The processes of certification and accreditation are discussed, as well as the need for effective auditing. The chapter ends with a discussion of the implications for their implementation in manufacturing and service sectors.

The text ends with an analysis of the future issues related to the development of integrated TQM.

Part 1

Quality Management — The Basics

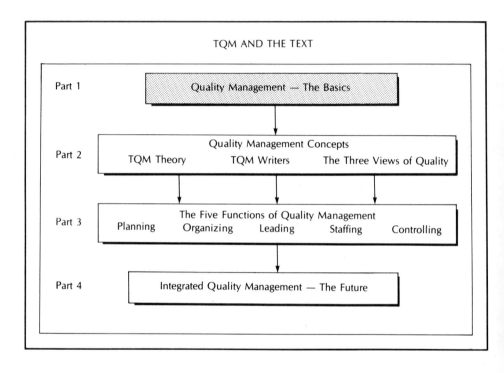

CHAPTER 1

The basics of management

CHAPTER OBJECTIVES

Define the term management
Evaluate the different levels of management
Explain what managers do
Describe the requirements of managerial skills
Evaluate managerial roles
Explain the development and contents of management theory
Discuss the functions of management
Explain the management process

CHAPTER OUTLINE

1.1 Introduction
1.2 Definitions
1.3 Levels of management
1.4 What managers do
1.5 Managerial skills
1.6 Managerial roles
    1.6.1 *Interpersonal roles*
    1.6.2 *Informational roles*
    1.6.3 *Decision-making roles*
1.7 The development of management theory
1.8 The theories of management
          ROBERT OWEN (1771–1858)
    1.8.1 *The classical theories of management*
          ADAM SMITH (1723–1790)
          CHARLES BABBAGE (1792–1871)
          FREDERICK TAYLOR (1856–1915)
          HENRY GANTT (1861–1919)
          THE GILBRETHS — FRANK (1868–1924)
            AND LILIAN (1878–1972)
          HENRY FAYOL (1841–1925)

               The functions of management
               Level of the use of the functions of management
               The management process
   1.8.2   *The behavioural theories of management*
               The pre-behavioural school theorists
                   Mary Parker Follett (1868–1933)
                   Hugo Münsterberg (1863–1916)
               The behavioural school
                   Elton Mayo (1880–1949)
   1.8.3   *The human relations theories of management*
                   Abraham Maslow (1908–1970)
                   Douglas McGregor (1906–1964)
   1.8.4   *The systems theories of management*
   1.8.5   *The contingency theories of management*
1.9   **Chapter review**
1.10   **Chapter questions**

1.1 Introduction

Management is a term that would seem to have been with us since the beginning of human endeavour. We seem to have been managing our lives, businesses, people, the way we do things for almost an eternity. But have we?

To provide a basis for the meanings that can be given to management, one or more definitions would seem to be needed. But these definitions are put forward not to be definitive, but just to illustrate its meaning. There will be other definitions that may suit differing circumstances, much like the application of the various theories of management. In developing these definitions, we must be careful to elicit a definition that will provide a 'wholeness' about the meaning of management and provide a more specific context in which that term 'management' may seem to reside.

1.2 Definitions

Definitions of management suggest the application of measures to ensure that financial, human and physical resources are planned, organized, lead, controlled and staffed effectively, or the achievement of goals through facilitating an effective process of planning, organizing, leading, controlling and staffing. These highlight the problems of attempting to define the term management.

    A set of activities (including planning and decision making, organising, leading and controlling) directed at an organisation's resources ..., with the aim of achieving organisational goals in an efficient and effective manner. (Griffin, 1993)

The process of getting activities completed efficiently with and through other people. (Robbins and Mukerij, 1990)

This emphasis regarding resources and people also indicates the varied ideas about management. It is therefore not surprising that managers themselves become confused about what they are supposed to be doing and what to do to make themselves more effective.

1.3 Levels of management

It is generally accepted that there are three levels of management — lower management, middle management and top management.

Lower management is normally concerned with directing or supervising staff in detailed, narrow task structures and processes. Consequently, these first-line management positions are not only a determinant of an effective organization, but a very important training ground for generating skills and experience of handling production people (for manufacturing or services) and possibly for managers further up in the organization.

Middle management generally directs other managers, not 'production' work people. In this instance, middle managers translate and implement top management policies and strategies and help to balance the pressures of top management with that of the capabilities of lower management in terms of work performance. There are fewer middle managers than lower managers. Some large organizations have a number of layers of middle management — true bureaucracies that have managers supervising managers. For example, General Motors has had at one time or another up to eighteen layers of management, with sixteen levels of middle management.

Top management creates policy, objectives and strategies that are used to guide the organization to achieve its aims. This is generally a small group of very experienced managers who have served in both lower and middle management capacities. Their terms of reference and their relative job contents also change from one organization to another.

1.4 What managers do

Henry Mintzberg (1975) made a very famous study of managers — although few in number — and found that their jobs could be characterized as:

1   *Managers work under constant pressure, at a unrelenting pace and on a wide variety of tasks.* This would seem to reflect the unorganized, frenzied approach given to management activities by many who are not trained in the *science* or the *art* of management. They also depict an ad hoc approach to solving management problems.

    Guest (1956) suggested that managers, in a working day, averaged 583 activities or an average of one per 48 seconds — quite a relentless pace. He did not however comment on the quality of work applied to these tasks or the effectiveness of their efforts.

2   *Managers prefer oral means of communications and networking.* Management behaviour revolves around the availability of information. Many did not have access to on-line information at the time they needed it and therefore resorted to verbal communications to effect this quickly. Today this may not be reflected to the same degree. In organizations with a quality orientation, where on-line information is practised, this has definitely changed, since there is shared power and responsibility through shared information.

3   *Management still seems to be more of an art than a science.* In the quality environment, managers will be expected, not only to understand and implement mathematical management techniques, but also to lead in its development in the organization. These are applicable to managing financial, physical and also human resources. These tools of the trade will allow managers to take their organizations towards gaining a competitive edge in the ever increasing world economy.

Thus, learning about quality provides the manager and subordinate with a more 'effective' management science, one firmly based on information.

Stewart (1976) collected data from hundreds of managers, using diary keeping, interviewing, observation and other techniques. For sheer coverage, Stewart certainly did a good job of uncovering some understanding of what managers actually do in real life. Stewart suggests that managers:

1   Do not operate along the lines of classical management theory (to be discussed later).
2   Are continually moving from one activity to another at a fast pace.
3   Communicate (oral and in writing) with other people more than they think for themselves.
4   Do not do the same things as each other and predominantly differ in their managerial attempts and aspirations.
5   React rather than being proactive most of their time, although this reflects the position in the hierarchy that the manager holds.

Kotter (1982) researched a group of general managers. He found they were not strategically oriented, proactive or well organized. This was surprising, as the textbook theory suggests that management as an art is learnt as they moved from lower management to top management. Instead of 'acting' the role of a top manager, they simply behaved in the role of lower managers, such as implementing decisions rather than making them. In this regard, general managers act and behave much the same as 'ordinary' managers. However, they do exhibit specific behaviours that separate them from the lower managers. Kotter suggests that they:

1   Engage in a range of activities that may be inconsistent with their position in the organization, i.e. they discuss concerns that should be expressed by lower managers.
2   Work long hours, through unplanned schedules, and react to other people's initiatives rather than developing their own.
3   Use *power* management, rather than coaching or other *softer* management options of gaining support through collaboration.
4   Develop networks as a means to effect decisions, but they do not make too many decisions in the first place.
5   Direct their management energies to issues outside their own undertakings and into areas that they forbid their subordinates to engage in, thus developing double standards that are inconsistent with their own behaviour and position.

1.5 Managerial skills

Katz (1974) identified four basic skills that managers use in the pursuit of their managerial endeavours. A skill is a learned ability to deal with a problem that can be repeated successfully, over time. Thus, the skill of riding a bicycle is learned through practice over time and can be repeated, even years later. These skills identified were:

1   *Technical* — Skills developed to effect proficiency in a given task, e.g. finance or production.
2   *Human* — Social and related skills that help a manager relate to others effectively, e.g. a manager motivating and communicating with subordinates.
3   *Conceptual* — The ability to evaluate, holistically and systematically, problems within and outside an organization, discern interrelationships and evaluate a balance of outcomes.
4   *Administrative* — The regulation of the activities under consideration. They may relate, to some degree, to conceptualization, but may not be catered for within a given situation.

The use of each of these skills within the organization differs for managers and is seen to depend upon the managerial level. Therefore, it would also seem to

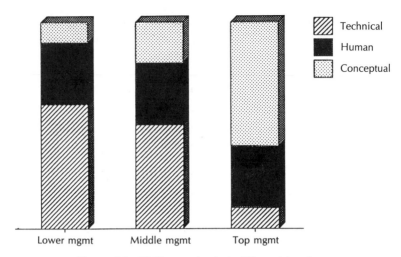

Figure 1.1 Skills required at different levels.

depend on the role the manager is allowed to play within the organization. Figure 1.1 illustrates the types of skills and the level found to be predominantly used or required within an organization.

Here, *Technical* skills are seen to be more important in lower management than at other management levels. *Human* skills have similar importance at all levels in the organization. However, human skills are required by all levels, because each level requires to communicate effectively, horizontally and vertically in the organization. *Conceptual* skills are very important at top management level, but not as important at middle or lower management. This reflects the decreasing detail managers work at, as they rise through an organization. Figure 1.1 depicts the possible relationship between these levels. These skills not only differ in response to size of the organization, but are also seen to differ in relation to manufacturing or service orientation.

Here, also, different skills can be seen in operation at the same levels within the same organization.

1.6 Managerial roles

As managers move up the hierarchy in an organization, they increase their control over the section or division of that part of the organization. Consequently, managers behave according to the *role* allocated to them through the selection processes. A role is a perceived, real or expected set of behaviours, determined

Table 1.1 Mintzberg's roles

Role	Function	Description
Interpersonal	Figurehead	Symbolic duties
	Leader	Enhances motivation of subordinates, coaching and communication
	Liaison	Development and maintenance of networks — both internal and external to organization
Informational	Monitor	Functioning as overseer and collector of relevant information that affect organization
	Disseminator	Distributes relevant objective and subjective information, gathered through official and unofficial means, internally
	Spokesperson	Distributes relevant objective and subjective information, gathered through official and unofficial means, externally
Decisional	Entrepreneur	Initiator and supporter of innovation and change programmes that affect the organization
	Disturbance handler	Effects corrective action when organization faces unexpected and unforeseen difficulties
	Resource allocator	Distributes financial, human and physical resources including time and equipment
	Negotiator	Represents the organization in important negotiations

to reflect a given position in an organization. Thus, the role of a top manager is different from the role of a workshop foreman.

Henry Mintzberg (1980) evaluated five chief executive officers. The extent of his studies suggested that many managerial jobs are not only similar in nature, but also in how the respective managers view their job roles. Mintzberg developed three functions — interpersonal, informational and decision-making — that have been divided into ten activities. These activities and functions are summarized here and can be seen in Table 1.1.

1.6.1 *Interpersonal roles*

The *Interpersonal* roles relate to the continuous application of routine managerial behaviours. They do this because they act as the central nervous system in the management and control of their unit areas. Consequently, the interpersonal roles are an important means by which managers deal with the workload. The *Figurehead* role is used to perform 'official' duties, such as representing the unit at formal meetings, both inside and outside the organization. The role of *Leader* provides direction, coaching, support and training for subordinates. The *Liaison* role is about dealing with people who make contact with the unit, both inside and outside the organization — customers, suppliers and even the competition.

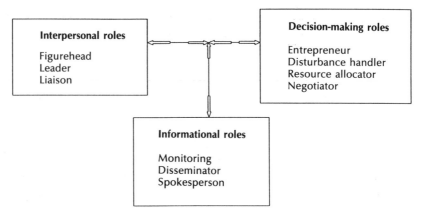

Figure 1.2 Mintzberg's roles.

1.6.2 *Informational roles*

The *Informational* roles relate specifically to the receiving and transmitting of information. This is the central purpose of the three identified functions (see Figure 1.2) without which no other role can be implemented. Most information that the manager has is gathered by oral discussion with subordinates or others.

The role of *Monitoring* thus provides a very necessary and important role. Constant monitoring provides the manager with information that can be used to make effective decisions, although the quality of the decision may bear no resemblance to the quality of the information received.

Another role is that of the *Disseminator*, where managers give subordinates information, either routinely or otherwise. The information may or may not be required by the subordinate, but may be useful to fulfil others requirements of social interaction.

In the role of the *Spokesperson*, information is given to people outside the organization. This may be through customers or agents or to the wider community through promotional activities. This role is underrated. Its importance is increasing, especially in top management positions, as a vital means to test the reality of organizational strategies that may have an ethical or social impact.

1.6.3 *Decision-making roles*

Managers depend on information to make decisions. Thus, one of the more important management activities is *Decision Making*. As entrepreneurs, managers initiate innovation and change to make their unit more operationally efficient and effective. The age of quality is transforming working practices, such that what were previously management activities are being given to shop-floor workers, a practice that is seen as flattening the hierarchy of an organization and reducing the bureaucratic jungle, moving decision-making activities to where commitment can be achieved and the outcomes can be implemented.

In the role of *Disturbance Handler*, managers respond to non-routine problems, such as strikes and major breakdowns. Here, procedures have not been effectively developed to deal with the problem and the experience, knowledge and skills of the manager are needed to bear on solving the problem.

In the role of *Resources Allocator*, managers have to balance the competing requirements of their own staff and physical and financial resources at their disposal. This also includes the time resource for both staff and the respective manager. Related to this aspect is the role of *Negotiator*. Here, competing demands whilst needing to be balanced, may sometimes cause problems, and negotiating may offer a means in which to change the scope of these balancing activities.

1.7 The development of management theory

We have previously discussed management and its meaning, the manager's role and what managers do. This is primarily a modern viewpoint, so how did we get to this point? Here, we will focus on the development of general management theory as a background to understanding how quality management has developed.

What, you may ask, is a theory? It can be seen as a tested hypothesis about the relationship between two or more observable, unobservable, objective or subjective events. It may or may not apply in all circumstances. If it applies in all circumstances, it becomes a *law*. If it applies in most circumstances, it becomes a *principle*. We therefore have a sort of loose hierarchy in which the hypothesis is the lowest and the law is the highest. Management theory can be seen to apply in all but the law — the highest.

However, not to get bogged down with this academic methodology, the application of either objective or subjective analysis does not mean the superiority of either, but the application of methodologies that provide a particular — but different — outcome. Consequently, the application of any management theory in any given circumstances may lead to the right or wrong conclusion. Since there is no unified theory of management, care is required when applying managerial theories to individual situations, because the outcome may not be as expected.

How did management theory develop? Management theory developed, not as a determined building pattern, but more on an ad hoc basis. This means that the theory has developed as a consequence of the failure in the application and assessment of previous management theories. New situations, new technologies and new reactions by people provided the positive or negative reactions to the construction and application. Modern management theorists tend to systemize management theory and provide a sort of unified development of the theory. An example of this is the management trends ascertained by Hickman and Silva (1987). Here, they discuss the six eras of management and determine that previous management eras are not forgotten, but added to. This would seem to suggest that there was a smooth development of management theory throughout this century. As we will see, many management theorists have been misinterpreted and the application of the management theory has been determined largely by capital owners and powerful managers who use the parts of the theory that can provide the most useful benefits to them. Consequently, this has supported the ad hoc view and is seen as far back as the 'first era' of management — the classical management theories — and it is there, that we now turn.

1.8 The theories of management

Management theorists have postulated for a very long time that management has developed through phases called management 'schools'. These have been determined as separate stages in the development of management theory, but only cover the last 100 years.

In general, the various different but related schools of management thought reflect the ideas and suppositions of their founders. In turn, this also represents the proposed solutions to many problems found at the time the theories were first developed. So we also have an environmental influence, and one that should not be forgotten as we explore and discuss them. What happened prior to this is largely ignored. Although management theory can be tracked as far back as 3000 BC, through many civilizations such as the Egyptians and Romans, it was not assessed as having a continuous development of a body of knowledge until the Industrial Revolution. But if we look at the machinations of the Egyptians (Burke, 1978), we find that the civil model, with the Head of State at the top, served by cabinet advisors and aided by a civil service is much akin to today's Western society. They had developed a very sophisticated society, using engineering, machinery, developing mathematics, law and the wheel to enact changes that assisted their society and provides benefits today.

Prior to the Industrial Revolution (c. 1750), work was generally carried out by craftsmen — who were generalists in the widest sense — and became known as the cottage system. They designed and built to order one-off constructions. They carried out all tasks necessary to produce their fabrication. They taught individuals through an apprentice system, so that they would eventually replace them. The use of power sources other than human or animal labour was minimal.

After the Revolution, the generalist concept was eroded and specializations began to take form — the factory system which we are all well aware of today. The development was enhanced with the invention of tools and machines and a new power source — the steam engine.

ROBERT OWEN (1771–1858)
Owen concluded that the Industrial Revolution was not only technological but

sociological in nature and, instead of freeing workers, the new economic system had created slavery. This is rather emphasized later in the behavioural school of management. Each school of thought was developed under prevailing social and economic changes and contributed to the evolution of ideas and strategies within that era. However, our discussion will focus on the developments accepted to have led to the age of quality today, and we start with the classical school of management theory.

1.8.1 *The classical theories of management*

Morgan (1986) discusses a machine metaphor, that organizations can be viewed as machines and have an input–process–output relationship. In this respect, this embraces the concepts and ideas developed in the classical school of management. Here too, Burns and Stalker (1961) discussed two aspects of organizations — one mechanistic and the other organistic. The mechanistic model is seen to be represented here.

The classical theories are predisposed to the clarification and development of the management functions and management principles. The writers at the time, for example, were trying to develop a more unified theory of management, but with a backdrop of increasing economic gains and ever-increasing pressures for organizations, such as manufacturers, to produce more goods. It eventually became a seller's market.

The world economy was booming and generally any production piece could be sold. Thus, outside environmental influences were not totally considered in the management of these enterprises. The focus was therefore on efficiency pure and simple. The human factor did not enter the equation either. Humans were expendable. They were seen as necessary in the production tasking, but were perceived as subordinate to the machine. As the machine's efficiency could be controlled, so was the worker's input closely controlled.

ADAM SMITH (1723–1790)
Smith is regarded as the forerunner to the first classical school — scientific management. He argued for a division of labour, which saved time, increased dexterity, and made the task of work easier, especially with the increasing development of newer technologies.

CHARLES BABBAGE (1792–1871)
Babbage is credited for writing the first management treatise in the machine age. He developed the principle of the transfer of skill, which refers to the increased deskilling of a worker, as the machine increases in automation.

FREDERICK TAYLOR (1856–1915)
Taylor was concerned with production. He was concerned with the efficiency of operations and this included the human aspect. Here, he focused on the worker,

Table 1.2 Taylor's worker principles

1. Develop a science for each man's work — the one best way
2. Scientifically select the one best man for the job and train him in the procedures he is expected to follow
3. Co-operate with the men to ensure that the work is in fact done in the way prescribed. This should include, but not be limited to, providing for increased earnings by those who follow the prescribed way most closely
4. Divide the work so that activities such as planning, organizing and controlling are the prime responsibility of management rather than the individual worker

Table 1.3 Normative principles of Taylorism

Always securing the full support of management

A complete mental revolution on the part of both management and workers

Workers should help management to establish scientifically the facts about production

Workers should agree to be trained in and follow new methods prescribed

Management should set up a suitable organization which would take all responsibility from the worker except that of actual performance of the jobs

Management should agree to be governed by the science developed for each operation and by the facts and in so doing surrender its arbitrary power over the workers

on the detailed operations that they were expected to carry out. To some extent, specialization must have occurred here, for him to make reference to it. He developed a large number of principles, some of the more important, in relation to quality management, can be seen in Table 1.2. These were first published in 1911. The basic normative principles that are quoted can be seen in Table 1.3.

Taylor was frequently described as an efficiency expert, and believed that by studying scientifically the specific motions that made up the total job, a more rational, objective and effective method of performing the job could be determined. Essentially *Taylorism* was about rationalizing jobs and making them more efficient and focused.

He described scientific management as a complete mental revolution on the part of management *and* workers. He stated that 'the great revolution that takes place is in the mental attitude ... under scientific management it is that both sides take their eyes off the division of the surplus, as the all important matter, and together turn their attention towards increasing the size of the surplus'. This implied a greater team approach than that for which he has been given credit.

Management's attitude was to ignore workers and use and develop the technology that was available. This strategy was based on the understanding at that time that machines could be developed to do the work of people, more consistently, and without tiring. Thus, efficiency became the goal, as it was assumed that people would respond in a rational, logical way and be motivated by economic incentives. Remember, at the time of the development of these principles, the economy was booming, all that could be produced could be sold, but the relative remuneration of the workers was small.

Thus, there was a predisposition towards incentives, but this was a management decision, not Taylor's. Under Taylor's system, people were viewed as extensions of machines. It was the machine's performance that dictated how fast a worker could be expected to work. Taylor had a concern for people that was largely ignored by management, who used his techniques to increase productivity and profits at the expense of working conditions and increased wages. Scientific management identified the 'human side' of organizations, but did not develop its huge potential.

Ford in 1914, at the Ford Highland Park Plant, introduced what is now known as Fordism. It is suggested that Fordism was actually a form of paternalistic Taylorism. This had provided a means by which increased efficiency, profits and quality could be realized, but at the expense of worker control. With the introduction of the $5 a day wage, Ford changed this. Workers were seen at this time to be driven by high wages and this provided the potential motivation to work at the plant. All workers are considered to submit to varying degrees of loss of control, whilst receiving remuneration in return. The capitalist viewpoint has always argued that if they pay your wages then you *should* accept their dominance in the workplace.

Ford believed that he had brought a sort of social revolution to the working practices at the plant and also to the consumer — who now paid a lot less for a car than under the old factory systems — from a price of $780 in 1910, to $360 in 1914. The use of Taylorism on the assembly line at this plant alone reduced the time to make a car from 728 hours — with a stationary chassis assembly and one man's work — to 93 minutes. Thus, the efficiency application of Taylorism outswept the human aspects.

What is so different today is that the age of quality can redress the balance between worker subordination and worker autonomy while still increasing efficiency. This brings up the notion that the present age of quality is actually Taylorism in disguise.

The principles of Taylorism, contained in Table 1.2, can also be seen as the underlying broad principles of quality management. However, Taylor wrote these principles after thirty years of work and at the turn of the century — when quality had not even been invented. This Taylorism extract is fundamental to human endeavour within quality directions. The quality directions of the turn of the century required the production of goods at a reasonable price, which performed to expectations. The same applies today; very little in this respect has changed.

This concept differs from Friedman (1977), in that he believes that a Taylorist strategy of direct control may be appropriate for semi- or unskilled workers, but that a strategy of responsible autonomy may be appropriate for skilled workers; the concept of quality circles rears its head. It is suggested that Friedman and the previous model are not mutually exclusive, but can exist in a 'happy' medium.

Since quality circles are developed and used to upskill workers, then this form of 'management' has some merit. Although workers still follow rules and procedures, they designed and developed those themselves with the help of management. The workers have thus become more skilled and Friedman's criteria would seem to have been met.

The quality drive provides the force for empowerment of workers so that, once again, they appear to gain control of their own work practices, while increasing productivity. For example, the use of quality circles provides the humanistic approach to develop the efficiency of an organization. Or is it that quality implementation can be seen as a means of introducing a **total** systematic culture?

This is much akin to ensuring standardization, reflecting **and** requiring worker subordination to set and operate by those developed procedures, whilst giving the *illusion* that they are in fact being empowered. Quality in this case is a furtherance of Taylorism at the administrative level. Is there any difference between Taylorism and this age of quality? Taylorism is still with us, and painful to some, but its impact is possibly increasing, not decreasing.

HENRY GANTT (1861–1919)

Gantt is renown for his development of a graphical chart for visualizing and scheduling performances. He was disenchanted with the use of Tayloristic principles, by the seemingly unscrupulous behaviour of management as an oppressive instrument. Consequently, he developed a system of management or working that replaced Taylor's principle of 'one best way' to 'best known way'. Thus, he attempted to humanize the scientific approach by ensuring that much less detail of the tasks contained in a job was required. It was still Taylorism, but performed at a more administrative level, especially since he reduced the conflict between management and workers, when he modified the pay system from a piece-rate system to a daily wage, plus bonus.

Again, what we find here is that management and management theorists have used the tools of this era, in this case the *Gantt* chart, and have divorced themselves from the humanizing element of the Tayloristic outcomes.

THE GILBRETHS — FRANK (1868–1924) AND LILIAN (1878–1972)

Frank developed Taylorism further. He carried out time and motion studies in the building industry. He was interested in developing the 'one best way' to do a job. Thus, he was primarily interested in efficiency and tried to develop ways to reduce job fatigue, in order to increase performance via increased output. He was a Taylorist at heart and used his time and effort to develop a system that reflected this.

Lilian however, was more interested in people. She was a psychologist and used this interest to study people at work. Lilian argued that scientific management should be used to develop the skills and abilities of people to their full potential (Wren, 1979). To her, scientific management was a tool to be used to the betterment of workers and consumers alike, not just for consumers.

HENRY FAYOL (1841–1925)

A mining engineer, Fayol considered administration problems rather than management problems. However, we can accept today that both are similar in orientation. He developed the concept of the universality of management and that management can be applied in all human endeavours. It has long since been accepted that Fayol was the founder of the classical management school because he simply systemized management theory and provided a workable prescription.

Taylor's emphasis was on the individual worker and the detailed tasks that they had to perform. Fayol was concerned with a more prescriptive management approach. Although they were contemporaries, when we analyze the output and strategies of what they wanted to achieve, they worked in opposite directions. Fayol worked from top management down, whilst Taylor worked from the bottom of the organization up.

Fayol developed what he called *principles* of management (Coubrough, 1930), many of which we recognize today being used in organizations. These principles are contained in Table 1.4.

Fayol had suggested that managers use five basic functions of management. He called these:

Planning, Organizing, Leading, Controlling and Staffing

The management cycle functions can be seen in Figure 1.3.

What is apparent here is that the principles that relate specifically to any 'human' connotations were *conveniently* ignored by managers and industrialists at the time. Moreover, they have been conveniently ignored to date by many management theorists. This is an appalling situation to reflect on, but one that is similar in context to the forgotten human elements of Taylorism.

Table 1.4 Fayol's principles of management

1	Division of work — The development of specialized work practices that would enhance the efficiency of the task
2	Authority and responsibility — Authority is the right to command and expect others to carry them out. However, this may not always occur, for varying reasons. Both authority and responsibility should be balanced
3	Discipline — This is an expected requirement of adhering to rules and codes. They should be fairly applied and understood by all
4	Unity of command — Only one superior for each employee because confusing signals and demands may develop
5	Unity of direction — Each employee, group and department should be working to satisfy the same aims and develop objectives that support these
6	Organizations should have a scalar chain, which is a chain of authority running from top to bottom
7	Equity — Employees should be treated fairly using published standards and requirements
8	Initiative — Worker initiative should be encouraged and should carry out plans for improvement
9	There should be a balance between centralization and decentralization

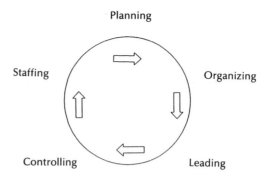

Figure 1.3 Management functions cycle.

The functions of management

It would seem that all managers practise these five functions in some way. More often than not, one or more of the functions become dominant, to the cost of the others. The job of management becomes more difficult and more complex, when we not only have to manage ourselves, but also others. In these circumstances their importance increases. In respect of quality management, this has become a very important issue, especially with regard to the increasing awareness of customers in expecting 'value for money' and generated higher expectations about the products and services provided and bought.

The functions of management provide a need for an integrated approach to management, much in the same way that 'quality' pressures have provided for an integrated approach in the design, development, production, selling and after-sales activities of the marketing and selling of products and services. They cannot be divorced from each other. Their importance and interdependence far outweigh the justification for ignoring and reducing the importance of any function.

In today's almost global economy — where there are increasing pressures of competition within many industries and commercial sectors — the effective application of the five functions of management has never been more necessary or decisive so, its effective application to quality management is still imperative.

What are the activities involved in each function? They include:

- *Planning* — This is the determination of targets and goals that need to be achieved, and using plans, procedures and strategies to achieve them.
- *Organizing* — Determining the right task to perform and ensuring it is distributed efficiently and effectively. It can be applied to individuals, groups, departments or whole organizations.
- *Leading* — Giving direction to others and getting others to do the task given. It means motivating subordinates and maintaining morale.
- *Controlling* — Determining appropriate standards, applying them and ensuring that they are achieved and taking corrective action when necessary.

- *Staffing* — Deciding the type of people to be employed and training them to do the jobs allocated.

In this text we are going to consider the application of the functions of management specifically from an organizational setting. This does not mean that these functions cannot be applied to the individual, but for the purposes of this book we will focus our discussions on the organization.

Level of the use of the functions of management

Does the application of the management functions vary with the level of management in an organization? Yes and no. For example, many managers plan, but the timeframe and implications of those planning activities vary in the organization.

The exact functional requirements of each job role will determine the dominant management functions and therefore the impacts of those functions. Other impacts include the organizational structure, culture and the orientation of the employees. However, it must be stated that every individual uses the functions to some degree in the organization, whether it is a private concern, semi-private or in public ownership.

Figure 1.4 illustrates the results of a study that provides for the general distinction between the various levels of management and the use of each function, based on time spent on each function.

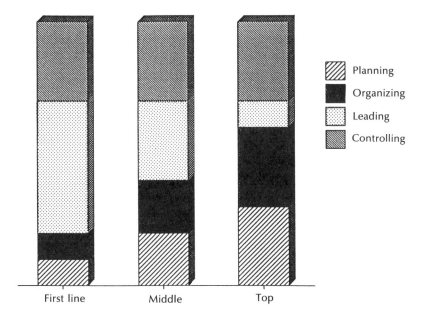

Figure 1.4 Management functions at different hierarchical levels.

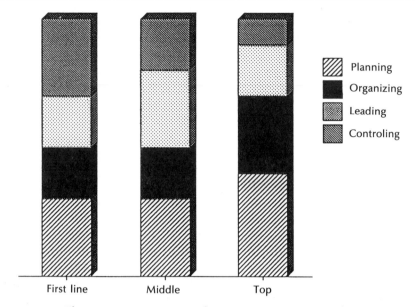

Figure 1.5 Management functions at Dynamics Ltd.

A different picture develops, however, when we explore a quality-oriented production unit. In this instance, Dynamics Ltd (James, 1994) provided the basis for a study of the use made of the five functions. The study determined that the top management functions, previously accepted, were actually being applied at a much lower level — as indicated in Figure 1.5, e.g. planning activities of lower management were increased by 200 per cent, and organizing activities increased by 150 per cent.

The imposing 'Age of Quality' has therefore provided changes to the management structures of organizations and will undoubtedly dictate further changes as they develop. This brings about the notion that the use of the five functions of management is a function of the tasks involved and the multi-skilling required to attain the necessary efficiency and effectiveness of the organization. Management practices have changed or will change as a result of an increasing quality orientation, and these changes will bring about greater self-management and autonomy for lower management and workers in an organization.

The management process

The management functions are not an isolated process. They are part of an interdependent system of management interactions. A model developed by Carrol and Gillen (1987) provides a basis for the additional elements in the management process. This model has been adapted further and is illustrated in Figure 1.6.

It is clear here that other interactions clutter the application of these

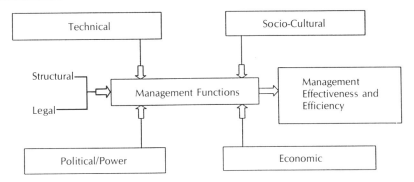

Figure 1.6 The management process.

functions in their purest form. An explanation of the various requirements may be necessary.

- The *technical* requirements include product, plant and equipment, investment, production controls, inventory and business processes, etc.
- The *political* requirements include external relationships of legal systems, governmental attitudes to business and internal implications of power.
- The *cultural* requirements include the values, beliefs, assumptions and behaviours of the various individuals and groups contained within the organizations.
- The *economic* requirements include the costs of goods and services produced, distribution and marketing, etc.

The result in real life is a balance between these dependent structures, since resources and time are scarce. So, management's function is to maintain a balance between the conflicting goals of these various elements. Management must also balance the pressures of the various publics contained in the management process — that of customers, other managers, other employees, directors, other stakeholders and so on. The internal environment of the organization is therefore as dynamic as the external environment and opportunities exist to provide meaningful development for managers.

1.8.2 *The behavioural theories of management*

The development of scientific management and its popularity with managers and capitalistic owners, signified the increased engineering and objective approach to management. As other disciplines became interested in learning how to manage, they provided differing perspectives that required different solutions. Gone was the accepted, clinical engineering approach as the only means to increase production, as more attention was being payed to the human element. Why was this?

Let us consider the economic situation in the middle to late 1920s. After the increased demand immediately before and after the First World War, there was a decrease in demand around this time throughout America and Europe. After the stock market crash of 1929, the economies of the Americas decreased considerably and therefore the markets for receiving produce. The market structure changed from being a seller's market to a buyer's market — very similar to today's macro-environment.

Consequently, the manufacturers had to improve their output without increasing the use of human resources. One way was to increase the use of automation, but this strategy required capitalization, which proved difficult to obtain. Automation also proved difficult to develop. The promise of automation — the production of goods without direct human endeavour — proved to be another false dawn in the consumer society. Another strategy was to develop the potential of the human element further. It was towards this strategy that the behavioural theories were heavily weighed.

The pre-behavioural school theorists

Behaviourists were involved in developing management theory long before the era of behavioural theories become noticed and the pendulum of change occurred. Accepted early behaviourists included Mary Parker Follett and Hugo Münsterberg. The discussion here is limited to these individuals in order to provide some basis for management theory development.

MARY PARKER FOLLETT (1868–1933)

Follett was exposed to classical management for most of her formative years, but saw the problems that it brought in the workforce because of her social-work interests. As a social worker Follett experienced, first hand, scientific management implications and as a consequence, developed a more behavioural, systems approach, in her management thinking. Follett focused on management as being a social process in which managers and workers operate. In this person we can see that she was not totally hindered by contemporary thinking, scientific management; but is this the case? Kast and Rosenzweig (1974) suggest that Follett was a major link between the classical and behaviourist movement. We could formulate an argument that the humanist approach that the behaviouralists presented was actually developed to some degree in the classical management era. Cognisance was therefore taken of these earlier ideas and repackaged in the light of the changing management environment.

Follett (1941) focused her attention firmly on the workings of groups. Here, her approach differed from classical management by being human-relations-oriented. Her contention, that groups influence how individuals work, was seen in practice later; in the Hawthorne studies. She determined that groups exercise control over themselves, what they do and want to achieve. Is this approach familiar? Quality circles, perhaps? Follett saw the benefits of people being involved

Table 1.5 Follet's principles of organizations

Co-ordination:
1 By direct contact. Here she emphasized that communication channels went both horizontally as well as vertically, as Fayol suggests
2 In early developments. Participation in policy forming activities will create commitment and motivation to carry them out
3 As a continuing process. Here she advocates a *power with*, rather than a *power over*. Co-ordinating inter- and intragroups was a social requirement as well as a management requirement

with the managing of the activities they participated in, was interested in the aspect of *power share*, where managers and workers should jointly develop together, rather than the now familiar coercive environment of manager against worker. However, she produced three co-ordination principles of organizations, which can be seen in Table 1.5.

This power share was the basis for resolving conflict through integration. Integration was used by Follett as a means for each party in a conflict to gain and not lose. Conflict resolution was a requirement of environmental influences and the equitable balance of seemingly opposing sides. Thus, Follett thought that working together provided the best means to effect changes — in response to changes. This was the human dimension — in a classical management environment.

HUGO MÜNSTERBERG (1863–1916)

Münsterberg argued that psychologists could use scientific management by determining ways that could help identify which individuals were suited for particular jobs. Also, they could help determine what conditions would provide the best means to increase worker performance. However, and possibly more sinisterly, they could help determine ways in which workers could be influenced to behave more appropriate to management requirements. So Münsterberg became known as the father of industrial psychology and as importantly, provided some further direction towards the behavioural school.

The behavioural school

ELTON MAYO (1880–1949)

Mayo was most remembered for his work at the now famous Hawthorne plant of Western Electric in Chicago. It was here that the behavioural school seemed to develop its roots. A number of studies were conducted by researchers from the National Research Council of the National Academy of Sciences in 1924, who considered the relationship between the physical environment and worker performance. This study considered illumination and the resulting worker performance. Two groups were selected, one to act as the control group and the other the experimental group. General lighting was increased and then decreased in the experimental group to see the production effect.

It was anticipated that production performance would rise with increased illumination and decrease with lowering illumination. In fact, production increased, even with a dimming light. In the control group, production also rose — and lighting had not been changed at all! The researchers retracted from further experimentation. What was the cause of this?

In 1927, Roethlisberger from Harvard Graduate School, under the direction of Elton Mayo, considered the relationship between conditions of work and worker performance at the plant. The results of the experiments conducted between 1927 and 1932 provided the framework of the development of the behavioural school.

One of the major results suggested that increased morale increases relative worker performance. Another result was that the experiments evaluated the impact of a movement away from individualism. The classical management school was supposed to be about individualism and the behaviour school was about the effects of collectivism. Since we already understand that this was Taylor's approach in theory, we also have the behavioural school actually repeating a theory from thirty years previously. The behavioural movement therefore determined that the individual was operating in a complex social environment with complex needs that must be satisfied. If these were satisfied then enhanced performance would result.

Brown (1967) summarized the results of Mayo, and these can be seen in Table 1.6. The major concept that resulted from these studies is the concept known as the *Hawthorne effect*. This refers to the change in positive attitude to their work of the workers — who are chosen for a study and raise their performance accordingly — rather than the specific factors tested in the study (Rice, 1982). However, critics have suggested that the concept is much too simplistic when considering the behavioural environment, which suggests that workers operate in a complex social environment. Thus, the Hawthorne effect may not be totally acceptable or realistic.

This does not detract from the achievement of Mayo, in that the focus of

Table 1.6 Mayo's Hawthorne studies — 1927–1932

1	Work was a group activity
2	The social world of the adult is primarily patterned about work activity
3	The need for recognition, security and sense of belonging is more important in determining workers' morale and productivity than the physical conditions under which they work
4	The worker is a person whose attitudes and effectiveness are conditioned by social demands from both inside and outside the work plant
5	Informal groups within the plant exercise strong social controls over the work habits and attitudes of the individual worker
6	Group collaboration does not occur by accident; it must be planned for and developed. If group collaboration is achieved, the work relations within a work plant may reach a cohesion which resists the disrupting effects of adaptive society

their studies was the individual, the group and the environment in which they worked, rather than the machine focus that seemed to have dictated many research studies in the classical school.

1.8.3 *The human relations theories of management*

An extension of the behavioural school was the human relations movement. Attention, given to workers, under the behavioural school banner, would raise worker performance. Co-operative approaches and developments were directed to enhancing the relationships between managers and workers.

The implications are that a social dimension has to be added to the managers and workers repertoire of skills, along with technical and conceptual — human skills (discussed previously). Thus the prompt is for managers to become more worker-oriented. Two management theorists considered to have developed the human relations school are Abraham Maslow and Douglas McGregor. Essentially, both determined different aspects of motivation theory — how a manager can motivate workers to perform better.

ABRAHAM MASLOW (1908–1970)
Maslow developed a theory of motivation based on two assumptions. First, human needs are insatiable and may or may not be consciously driven. Second, humans work to satisfy those needs. He developed a hierarchy of needs that provided some understanding of where an individual's needs are distributed. They range from the lowest need of physiological, through safety, social, esteem and to self-actualization. This hierarchy was an important development, because it highlighted the fact that humans had multi-level needs that could not be satisfied by just plain economics.

DOUGLAS McGREGOR (1906–1964)
McGregor developed a dichotomy, theory X versus theory Y. The assumptions of each are contained in Table 1.7.

These assumptions determine the impact of the attitude of the manager. If a manager uses theory X, then a more controlling non-trusting environment develops — much closer to the application of classical management theory. Whereas, if the manager believes in theory Y, then the environment will be more humanized and cognisance will be given to the human element. A more integrated approach will be used to create commitment to organizational goals. The theory X and Y approach has been used extensively since it was first mooted in the early 1960s.

Managers have found that the application of either theory must be dependent on the individuals that work for the manager and implies the application of personal knowledge on the preferences of the leadership-style requirements of that individual. Essentially, this means a sort of contingency approach — but more on this later.

Table 1.7 McGregor's theory X and Y

Theory X
1. The average person dislikes work and avoids it if they can
2. People need to be coerced, controlled, directed and threatened with punishment to get them to work to organizations' objectives and goals
3. The average person wants to be directed, shuns responsibility, has little ambition

Theory Y
1. The mental and physical effort required to work is as natural as play
2. People, when given the opportunity, will exercise self-direction and control to reach goals to which they are committed
3. Commitment to goals is a function of the rewards available, especially high rewards such as recognition
4. The average person seeks responsibility and has high ambition
5. Many people have the capacity to exercise a high degree of creativity and innovation in solving problems

Reprinted with permission from McGregor, *The Human Side of the Enterprise*, McGraw-Hill, 1960

1.8.4 The systems theories of management

A system is any unified entity, where all parts are interdependent and interrelated. Organizations can be viewed as such entities. Kast and Rosenzweig (1973) provide an example of the types of organization subsystems. These include technological, strategic, managerial, cultural and structural subsystems. The system can also be viewed as closed or open.

Most organizations are of the open variety and we will adopt this view here. Open systems can be viewed as an input—process—output configuration, both internally and to the external environment (see Morgan, 1986). Consequently, these organizational systems react with that external environment and are therefore not immune to external change.

An advocate of the systems approach is that of Churchman (1968). Here, Churchman takes a quantitative approach to systems analysis. He suggests that all systems have four characteristics, these being:

1. Environment — all systems operate within an external environment in which there are input and outputs.
2. All systems, by definition, comprise of interrelated parts.
3. All systems are characterized by their interrelatedness.
4. All systems have a central function or purpose.

To add to the above, a system must have some sort of boundary, rigid (closed system) or flexible (open system); develop synergy, which means that the whole is greater than the sum of its parts taken individually; have a flow of information, materials and energy; and have a feedback mechanism that is the key to system controls (Kast and Rosenzweig, 1972).

For example, to ensure that a given department is meshed into the operations of other departments in the organization, managers must communicate, enter into dialogue that includes information transmission and represent the department's interests to the other departments in the organization (Tilles, 1963).

1.8.5 *The contingency theories of management*

Contingency management is the application of a management style, depending upon what the situation demands. It requires a careful understanding of the needs of the task, people involved and the processes needed to satisfy the organizational goals. The contingency theory of management is a hybrid management style, one that is deemed to be flexible and responsive to the situational demands. It is therefore not surprising that this is also called the situational approach to managing. The theory was built out of the failed applications of the management theories previously developed, or more likely, the application of parts of the theories that appealed to the respective manager at the time. It is interesting to note, that the application of scientific management was an example of testing contingency management. The extraction of the scientific part of Taylorism, instead of also using the humanized element, undoubtedly brought about the ultimate dissatisfaction surrounding the theory. The major problem that exists here is that managers are people themselves; they adopt a management style that reflects their personal understanding of the situation and therefore may not quite read the situation appropriately or effectively.

The scientific-management-oriented manager may move predominantly for this school of management, not because he/she does not know the basis for other management schools, but because of trust in personal competencies in managing in any other style. Likewise, the psychologist may predominantly apply the behavioural school or in some cases scientific management. So what we have here is the application of the different theories to different situations, based on the flexibility of the preferred management style.

What is the focus of the contingency theory of management? Luthans (1973) has studied contingency management and suggests that contingency management uses systems theory as a base framework, determines the relationship between the parts of the system and defines the specific tasks that need to be carried out in order to enhance the efficiency and effectiveness of the working unit.

Critics of this theory suggest that the theory is not sufficiently developed to be called a true theory (Koontz, 1978). Other comments are that this theory adds nothing to the theory of management, but does allow for greater experimentation. Even Fayol suggested flexibility in the application of management practices and styles of leadership.

Proponents of the contingency approach view the problems and solutions need also to have flexible considerations, that may be different, in different circumstances. So the application of the right theory is viewed as insufficient. We have to ensure that the proposed solution can be equitably and effectively applied also.

Woodward (1965) researched the relationship between technology orientation, the task and the performance and effectiveness of employees that result. Her studies suggest that in some circumstances a flexible management system is more effective than one based solely on the classical management theories. Here, Woodward described a flexible management system as one that responds to the relationship demands between technology and human resource requirements.

Earlier, Burns and Stalker (1961) commented that there is no optimum type of management system and therefore the contingency approach will develop as perhaps the only means to deal effectively with the complex sociological interactions that occur in organizations. Contingency management is therefore a theory looking for an application, and supporters and other management researchers are continuously assessing its capabilities and development.

1.9 Chapter review

The development of management theory over the past 100 years or so has been discussed in this chapter. However, this only provides a very basic introduction to management theory.

The initial development and implementation of classical management theory was determined in an environment of growth. This signified the need, as ascertained by managers, to focus efforts on increasing efficiency through use of technology. This was not only to 'ensure' production, but through a backdrop of educational limitations of the workers. This is primarily an American view, but since the classical theories brought about a significant impact with Taylorism, this seems to be a reasonable bias.

Critics of scientific management, saying that it dehumanized jobs, were responding to the way managers worked and responded, which reflected the efficiency focus via technology development.

Even before the great crash of 1929, the human relations movement had been developing. It had a significant impact, through Mayo, on the general switch from applying classical theories to the behavioural theories. The application of classical theories were still applied, but more cognisance was given, for example, to the effect of technology on the human being.

Contingency theory, although not accepted as a management theory *per se*, has promised to provide us with a process means of dealing with the complex sociological environments contained within and outside organizations. The multiplicity of tasks will thus generate a multiplicity of management processes and styles to accommodate the expected changes in management thinking.

However, the movement to a multiplicity of diagnosis and application reflects, to some degree, other changes that have occurred in the greater social environment, which, by default organizations and individuals who work in them are part of.

1.10 Chapter questions

1. Define the term management and outline its importance.
2. Evaluate the effect of the different levels of management on both a large and small organization.
3. Explain what managers do.
4. Describe the requirements of managerial skills to an individual who has just been promoted to their first supervisor's position.
5. Evaluate the managerial roles found in an organization you have knowledge of.
6. Discuss the five functions of management.
7. Explain the management process.

Part 2

Quality Management Concepts

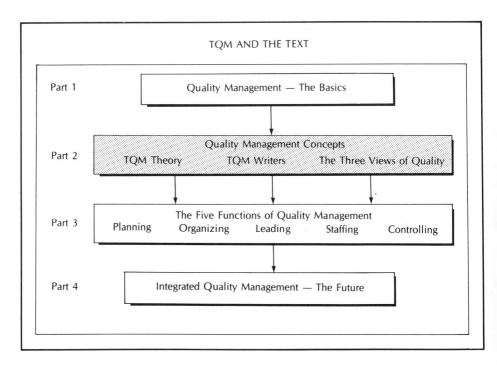

CHAPTER 2

Total quality management

CHAPTER OBJECTIVES

Discuss the development of TQM
Define the term TQM
Explain the basis for TQM
Evaluate the various issues surrounding the implementation of TQM
Discuss the process of implementation of TQM
Discuss the management functions and the issues to be considered when implementing TQM
Explain the realities associated with TQM in manufacturing and service sectors

CHAPTER OUTLINE

- **2.1** Introduction
- **2.2** The four eras of quality management
  - 2.2.1 *Quality development through inspection*
  - 2.2.2 *Quality development through quality control*
  - 2.2.3 *Quality development through quality assurance*
  - 2.2.4 *Quality development through TQM*
    - Implementation issues
      - *The organization*
      - *Management commitment*
      - *Culture change*
    - Management functions and the issues to be considered when implementing TQM
    - The process of the implementation of TQM
    - TQM practices — the reality
      - *Sector experiences*
- **2.3** Chapter review
- **2.4** Chapter questions

2.1 Introduction

The development of *quality* has been essentially continuous over the past 100 years. Although it must be said that quality existed before this time, its systematic interest and 'labelling' as quality has brought the changes in society we see at present. This development cannot be attributed to one person or a handful, but over the years the practical implementation of management strategies that sought to satisfy customers needs and wants has brought us to the present *age of quality*. Table 2.1 illustrates a model of the interaction of the eras of quality management and the main features that characterize that stage of development.

The aim here is not to distinguish — at least in the explanation of the development of TQM — between manufacturing and service requirements of quality, as this will be discussed later in the ensuing discussion of the implications of the five functions of quality management.

2.2 The four eras of quality management

2.2.1 *Quality development through inspection*

Garvin (1988) suggests that the development of quality management started with inspection, that the outcome of the Industrial Revolution was the development of specialists who 'inspected' quality into a product. This seems a reasonable attempt, but flawed. Scientific management, as previously discussed, did not happen because some managers decided it should. It occurred because of responses to environmental influences, both inside and outside the organization. Scientific management provided the backdrop for the development of quality management through inspection.

Let us go back in time to when craftsmen were the only producers of goods. If we look at their techniques, in simple terms we find that at every stage of the manufacture of the product inspections took place. They did not just happen at the end of a particular stage, they occurred at every portion of the product manufacture, that is, each component was continually inspected to ensure that

Table 2.1 Quality management theory development

Feature	The four quality management eras			
	Inspection	Quality control	Quality assurance	TQM
Pre-scientific management	*			
Scientific management	*	*	*	*
Behavioural management	*	*	*	*
Systems management			*	*
Cultural management				*

it was correct for the next stage. Where a flaw developed, the item was cast aside to be returned to the raw materials input, if possible. What we have then, through the craftsmen, is a process that looks much like what we are attempting to do today — quality management through inspection. However, the major difference between then and now is the number of items produced. Another major difference is the attitude of the craftsmen, who were individualistically oriented, rather than the team orientation as required under TQM.

Because craftsmen produced items by hand, they did not produce many, were slow to make changes, were process-oriented and not customer-oriented, and generally sold what they could produce, due to a production economy orientation. However, this changed with the development of technology. Craftsmen's guilds resisted the introduction of technology to assist with increased production requirements; this provided for a strategy of standardization. This proved problematic, because craftsmen were generally incapable of producing to specification every time. They produced items that generally reflected the specifications, but those were inconsistently determined. Consequently, standardization and craftsmanship were at two ends of the spectrum, and could not meet. Thus, with the requirements of higher production, standardization became the norm, rather than craftsmanship — and scientific management was born.

Scientific management required that each job was broken down to its smallest manageable component. This had the effect of deskilling the craftsman's job. The resultant breakdown of tasks enabled any individual who received a small amount of training to do the task. Craftsmen were therefore made redundant in many areas of production.

The problem that occurred was the divorce of responsibility from the actual outcome of the task, i.e. as long as the incumbent produced the item according to the task process specification, then inspection was not needed. Inspection of *finished* articles thus became the norm. The problem was that in-process equipment did not keep *in-process*, and the resultant defects were not accounted for until they were inspected at the end of the line. Specialized personnel, mechanics and other technicians, were employed to ensure the machinery and equipment were maintained effectively. This did not allow for process movements while operating the equipment and the resultant defects were only detected later. The added value given to the item with the defect is thus wasted and adds cost to the production process, and from there to the consumer price or reduced profits.

As the economies changed towards more competitive, discerning customers — at least in the United States — one major strategy that was adopted was the reduction in price to the consumer. This required cheaper input prices and process costs. To achieve this, emphasis was placed on making the production line more efficient, by reducing variable costs — such as humans — and automating the process as much as technology allowed. The Ford Highland Park plant was one such result.

Increased technology use in production allowed greater control over the standardization of the product produced. As a consequence, designs were achieved that allowed interchangeable parts. This created even more pressure towards standardization.

Significantly — in 1914 — the requirements for arms production, in both quantity and interchangeable parts, produced the impetus for the development of this much needed production technique. What was done to allow for the inspection of parts coming off the production line continuously? Some form of measurement system — jigs, fixtures and gauges (Hounshell, 1984) — was required. These were support tools that helped to manufacture products. The measurement devices were essential items that allowed a relatively quick analysis of the items produced. Where problems occurred in manufacture, specialists were called on to repair or replace the support tools. Since there was a lot of pressure because of short timeframes — because of production holdups — errors were frequent. Even in the best organizations, some items were defective. To ensure that this did not prevent a customer getting their product, large-scale inspections were required — every piece that was produced was inspected and either accepted or discharged. The power shift therefore, went from front-line staff — as in the craftsmen era — to the inspection staff, in this era of scientific management.

The use of gauges provided more objectivity to the determination of whether a product met the standards set in the production specification. Thus two or more inspectors could create a somewhat objective reality that further provided for the importance of the inspection teams themselves. This provided the immediate basis for the development of scientific management as developed by Taylorism, and one of the reasons why Taylorism, in the form of this objective reality, became so popular. Thus, 'stand-alone' inspection grew out of the necessity to produce interchangeable parts, and the basis of quality control created.

Radford (1922) discussed the link between inspection and quality control. He also determined the need to include designers in the quality development process, and the use of quality control in quality improvement with its associated increased production, lower costs and benefits to the consumer — and this was in 1922! This had been discussed earlier, in an article in 1917.

What is evident is that this was a major development in the quality field, but few people have taken any notice of it. What a pity. We might have been a further twenty years down the quality track by now. Radford was indeed ahead of his time when he further discussed that the purchaser's (consumer's) requirements are met when the manufacturer adheres to these requirements in the manufacturing process.

As befits the scientific management approach, emphasis was placed on simplifying the inspector's task, and it became obvious that inspection capabilities were limited, not by the ability of the individual, but by the capability of the tool used. This created a number of problems, because the counting and sorting of items produced were all that was generally required, a very erroneous task because of the specialization that created it.

2.2.2 Quality development through quality control

Managing quality through control means dealing with data, developed from the actual process used to produce products or services. Since the products or services are always produced to customer specifications, effective control of the manufacturing process will result in consistent and standardized output that will meet their requirements every time. It means less waste, more efficiency and probably greater profits.

Walter Shewhart, as far back as 1924, while working for Bell Telephone Laboratories in a department later to become a quality assurance department developed the concepts that are basic to statistical quality control. This was later developed into a formal text, which has become the change point in scientifically accepting the discipline of quality control.

For the purposes of this text, it will be assumed that this is the start of statistical quality control. 100 per cent inspection did not necessarily provide the best way of guaranteeing non-defective batches of product. It was also, possibly, time consuming, impractical and invariably expensive.

Shewhart first recognized that the principles and practices of probability analysis and statistics could be applied to the quality problems in manufacturing. He also recognized that the manufacturing process was variable in nature and that this variation occurred throughout the process and through time. This effectively meant that products could not be totally standardized in essence, but could consistently be produced within a given tolerance. Single products would vary, even in a single process with one machine and one operator working to a simple specification.

The other major concept was that no two processes were the same and that product needing to be matched at the end of each process would be subjected to the variation found in each. This produced wasted products. It was just this that forced engineers and designers to resort to scientific management earlier and use less sophisticated machines to produce much less sophisticated products. Thus, the problem of variation had been tackled, unknowingly, by the very engineers and designers developing production processes in the scientific management era.

Management now needed to determine what variation was acceptable and what not. This then brought the question of how to make this assertion. Shewhart applied simple statistical techniques, such as the $\bar{X}$ and R charts (see Chapter 13). This allowed assignable (variation that was not normal) and non-assignable causes of variation. The idea was to segregate chance and actual (real) causes of variation and manage them effectively. If this simple quality control process were implemented at each stage in the manufacture of a product and when a process exhibited assignable causes of variation and was corrected, then this saved time and money (adding value to a product) both on the product to that point, and further along the production process. This way, control of the production process was effected, and as a result, control established over the consistency of the product.

The major difference between inspection and control was the focus. In the inspection era it was on the product, and in the quality control era it was on the process. Unfortunately something else was being developed at the same laboratory — sampling. The consequence of Dodge and Romig was to have a critical and perhaps an inopportune effect on the implementation of quality control.

Initially, sampling provided the basis for end-of-line inspection, rather than continuous monitoring. This rather limited the use of charts as a method of managing the production process. It defeated the advantage of charts by preventing internal waste and allowing continuous management of the production process at *all* production points rather than just at the end. Unfortunately, sampling also provided a means by which management could practise scientific management, that is, by training a few people who could be trusted by management to determine the effectiveness of a production line. Management generally still did not trust the shop-floor workers and worse still, did not view many of them capable of excelling at their job in the way management wanted them to. Consequently, sampling developed into a specialist job — but not the one intended by Shewhart.

Sampling had its limited advantages though. It could provide an effective and efficient way to analyze a batch of product and determine whether it should be shipped to the customer. Sampling could assist where destructible testing was a requirement — as in mineshaft ropes. It meant examining a small number of output products and determining their state of acceptance or non-acceptance (more on this in Chapter 13).

It would seem that quality control in this form had become the norm as the Second World War came and went. There would be few changes until the early 1960s. Why was this? If we examine the economic environment during the mid 1930s to mid 1950s, we generally find that the economy was once again a sort of command economy — one where whatever was produced was bought. This will explain why sampling became an end-of-line activity, rather than being accepted as a production-point activity. Whatever was produced would certainly be bought by the government or the private consumer. It was also during these times that America got fat on waste, because no one worried about the costs of waste — companies were making enough profit to cover those.

2.2.3 *Quality development through quality assurance*

Quality assurance saw, in Garvin's (1988) terms, an evolutionary tract that took quality from a narrow perspective — where it was purely the domain of specialists — to much more broader perspectives which involved management to a greater extent. No longer was the differentiation and specialization of jobs effective. Increased awareness of the implications of quality throughout the whole workforce, management and of course through to the customer was now required.

Quality started to become more than a specialist concern. Improvements

Table 2.2 Quality management eras and their foci

Era	Focus
Inspection	Product
Control	Process
Quality assurance	System
Total quality management	People

in quality could not be brought about without commitment from the people working on the shop floor. This increased the focus of the quality revolution. Management determined that quality could be 'assured' at the place of manufacture.

BS EN ISO 9000 or BS 5750 states that quality assurance is all 'those planned and systematic actions necessary to provide adequate confidence that a product or service will satisfy given requirements for quality'. Quality assurance needs to use quality audits in this respect. Quality audits are designed to provide real evidence of the integrity of the production system through an independent examination. Thus, quality assurance is about developing an internal system that develops data over time that would signify that the product produced was to the specifications and that any errors were detected and removed from the system. This then provided the basis for the production system improvement cycle we see so commonly today. Quality assurance systems can now be seen implemented in many departments in an organization, for example, marketing, production, finance and supplies. Each system is self-standing and at times independent of the others. This requires close co-ordination between the departments, which sometimes breaks down.

The quality management eras seem to provide a basis for the continuous and discrete development from one management period to another. However, the focus of each era has been different. This is highlighted in Table 2.2.

The latest era is TQM, and it is here that we now concentrate.

2.2.4 *Quality development through TQM*

A model of total quality management (TQM) can be seen in Figure 2.1. This model provides the basis for the following discussion.

Atkinson (1990) said 'TQM is an organisation-wide commitment to getting things right'. TQM affects everyone in an organization, and it is thought that in order for the organization to be competitive and therefore successful, the philosophies, principles and practices of TQM must be accepted by everyone. Oakland (1989) suggests that TQM is an 'approach to improving the effectiveness and flexibility of business as a whole'. Essentially, TQM requires a revolution — a cultural revolution — in the way people do things in an organization.

The model contained in Figure 2.1 illustrates the areas that are considered necessary for the constituent parts of TQM. The mix of such elements is dependent upon the circumstances facing an organization — both internal and external. This

Figure 2.1 TQM influences.

would imply that each element must be readily seen in an organization, but it does not help us determine the appropriate portions of each element or the emphasis to be given.

Although it can be assumed that the elements are interdependent in some way, they must exist to some degree to ensure TQM philosophies and practices which therefore support and help develop the TQM culture. TQM can be described as the management philosophy that seeks continuous improvement in the quality of performance of all processes, products and/or services of an organization. Atkinson (1990) states that 'Total quality is a strategic approach to producing the best product and service possible — through constant innovation'. TQM offers a means by which organizations can provide employee participation, customer satisfaction and, just as important, organizational competitiveness. It emphasizes the understanding of variation, the importance of measurement and diagnosis, the role of the customer and the involvement of employees — at all levels of an organization — in the pursuit of continuous improvement.

The successful implementation of TQM generally requires the use of specialist knowledge. These experts may be quality standards auditors for setting up standards and work practices, or change masters, to effect the culture change of the organization. Nevertheless, if top management is committed to the quality cause, then implementation using specialists, although expensive initially, will pay dividends in the long run. Lower overall costs are likely, worker expectations are fulfilled and customer loyalty and satisfaction generated. Often, because of politics, it is possibly better for top management to be seen to support the changes, but not to be instrumental in the change itself. Middle management, in particular, may recognize the implications change may have for their job security and as such not co-operate with somebody they feel is not supporting them.

TQM is not a panacea for all the productivity ills that can sweep organizations, but can provide the means by which the essential change patterns of modern business can be grasped and redirected, to provide opportunities never thought possible for an organization without the quality vision. TQM is a philosophy of management, generated through a practical orientation, devising a process that visibly illustrates their commitment to growth and even organizational survival. It means focused action, leading to improved quality of work and improving the organization as a whole. It enables an organization, through a co-ordinated strategy of teamwork and innovation to satisfy customer expectations, needs and requirements. The problems raised by implementing TQM are considerable and need to be addressed as early as possible in the change programme.

TQM demands:

1. Visible organizational values, principles and standards, which must be accepted by everyone.
2. A clear strategic business orientation — mission, quality policy and quality objectives — with practical and effective procedures and practices.
3. Clearly developed customer/supplier (internal/external) requirements.
4. Demonstrated ownership of all processes and their relative problems.

TQM requires the development and application of education and training programmes for effective business management, knowledge and practices of specific tools/techniques, which enable continuous business improvements to be made. Deming (1982) uses the Shewhart cycle to illustrate continuous improvement. This has now become the PLAN, DO, STUDY, ACT (Deming, 1993) cycle shown in Figure 2.2.

> In the last decade many Western organizations have come to appreciate the strategic importance of total quality management (TQM) to their corporate health. They have realized that TQM will enable them to become and remain competitive in home and international markets. (Dale and Plunkett, 1991)

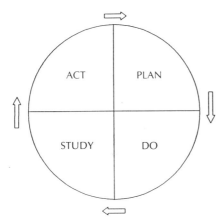

Figure 2.2 The Deming/Shewhart cycle. Printed with permission from Deming, *Quality, Productivity and Competitive Position*, MIT, 1982.

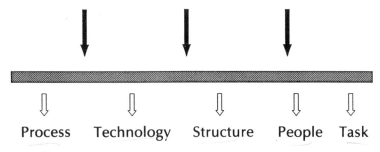

Figure 2.3 Five systems of total quality management.

TQM has been given many definitions, as we have already encountered. However, there exist common threads between many of them, such as the definition provided by the British Standards Institution (BSI) — BS 7850 (discussed further in Chapter 15). What is surprising is the fact that the Japanese apparently have a dislike for the term, preferring instead to refer to company-wide quality control or total quality control (Ishikawa, 1985).

TQM is deemed to require five system elements. These are illustrated in Figure 2.3, they are technology, structure, people and task. Without the effective balancing of these systems, TQM loses much of its power to effect change. *Process* includes managerial, administrative and production processes. *Technology* includes any items, components or articles necessary for the task accomplishment. *Structure* includes individual responsibilities, formal segmentation conditions of the organization and formal and informal communication channels. *People* includes education and training, culture change, etc. *Task* includes quality issues, job functions, etc. These are illustrated in Figure 2.4.

Implementation issues

There are several major steps to be considered before the implementation of TQM. Some of these require an attitude of commitment and co-operation — specifically by top management, and a willingness to experiment and be flexible by everyone

Process	Technology	Structure	People	Task
Organization and systems	Production line	Responsibilities	Team building	Quality issues
Quality planning	Information use	Communication	Education and training	Culture change
Organization Leading Controlling		Administration	Management Development Rewards and reinforcement	Job Functions
Design methodology Auditing				

Figure 2.4 Five systems of total quality management explained.

TOTAL QUALITY MANAGEMENT

in the organization. If these are not present, then the top priority of management must be to create a cultural change towards the development of these positive attitudes. There is the requirement of an enhanced commitment of top management, not just rhetoric but in practice.

In addition to the development of positive attitudes and TQM awareness, a very good understanding of the principles and practices of TQM is required. Oakland (1989) indicates that the '*preliminary stages of understanding and commitment are vital first steps . . . The understanding must be translated into commitment, policies, plans and actions for TQM to germinate*' (author's emphasis).

Some issues seem more critical than others for the successful implementation of TQM. The following issues seem appropriate for discussion.

The organization

The need for TQM to be adopted organization-wide is seen as paramount. Oakland (1989) states that 'TQM is an approach to improving the effectiveness and flexibility of businesses as a whole. It is essentially a way of organizing and involving the whole organization'. Trying to implement TQM will mean that a plan will have to be developed considering the whole organization.

To limit TQM to one department will mean that co-ordinating tasks between different departments may become more complicated. For example, the process needs of a department operating with TQM will be different to that operating in a more conventional management orientation. Communication may therefore suffer. An attitude of *them and us* could be created, working adversely to the concept of TQM, which is centred around teamwork. Piecemeal applications of TQM should not be indulged or considered.

Management commitment

The acceptance of the need for TQM and with it the commitment required to drive the new management system, particularly in the initial stages, will be of prime importance. Again, the importance of the attitude and commitment cannot be overemphasized. Whereas we may succeed without proper plans, we will never succeed without commitment and the development of a positive attitude. Commitment by management at the highest levels in the organization must be genuine and visible. Oakland (1989) indicates that 'to be successful in promoting business efficiency and effectiveness, TQM must be truly company wide and it must start at the top with the Chief Executive, or equivalent, the most senior directors, and management, who must all demonstrate that they are serious about quality'.

The positive attitude must therefore be considered as the first and encompassing issue when considering a change to TQM. This means that the underlying culture of the organization must possess the attitude to accept this change willingly and commit itself to it.

Culture change

The need for the culture to change to one which values teamwork and flexibility is paramount, the realization by staff that by belonging to a team and making worthwhile team efforts, working conditions, staff and customer satisfaction will be enhanced.

Atkinson (1990) makes the following statement about quality and attitude:

> Quality is about attitude. Spending money does not promote quality, although it may provide the tools. Throwing money at problems does not make them go away.
> Investing in technology and robotics is not a route to TQM. You may invest in the wrong technology or implement it badly. Quality is something which is engineered through effective human relations.

In this statement, the attitude is an integral part of the new culture. Regretfully, the existing cultures in many organizations do not have the basis for the attitude required for successful implementation of TQM and it will be necessary to evaluate the culture in light of the projected changes. If found to be inappropriate, a cultural change needs to be engendered through induction, education and training of all personnel, to ensure that the attitude and acceptance of ownership of and commitment to TQM will be forged.

The resistance to TQM is partly due to unfamiliarity with the new approach and how it will affect the staff in their jobs. Loss of control, having to do things differently and the uncertainty of being able to cope with the new demands of the altered job description, can create suspicion of the change, which needs to be overcome. In general — especially in the initial steps — it is also perceived that more work will be created.

Management, in the eyes of employees, may not have a history of success with implementation of earlier initiatives. It is important to consider the impact of failed past ventures. Is TQM regarded as another fad of management which, if left to its own devices, may disappear in the next few months? We have tried something like this last year and it also failed; or, 1993 was the year for teamwork, 1994 is TQM, what is new? Whatever the type of intervention, there is a need to focus on all members of the group, and not only on a few. This corresponds to the system-wide intervention of most organizational development programmes.

Atkinson (1990) states that 'there is little point talking about the joint approach to implementing company-wide improvement through TQM, when little effort is made to ensure that the drive is based on company-wide participation, where it is discouraged by strict demarcation between functions and departments'. This seems to be an accurate description of the state of many organizations prior to quality management intervention. It is therefore necessary to ensure that TQM is communicated to and ownership of its role taken on by all staff, so that a team effort or company-wide commitment will be ensured. Ongoing training is needed, specifically of new entrants, so that the company culture is present from the outset. It should not be forgotten that staff need to remain motivated through the actions of management.

The team approach is necessary so that open communication and joint problem solving may be used. Oakland (1989) indicates that

> the use of the team approach to problem solving has many advantages over allowing individuals to work separately on problems:
>
> 1   a greater variety of problems may be tackled, which are beyond the capability of any one individual, or even one department
> 2   the problem is exposed to a greater diversity of knowledge, skill and experience
> 3   the approach is more satisfying to team members and boosts morale
> 4   problems which cross departmental or functional boundaries can be dealt with more easily
> 5   the recommendations are more likely to be implemented than individual suggestions

Management functions and the issues to be considered when implementing TQM

Figure 2.5 indicates some of the many issues considered to be involved when implementing TQM. These issues include:

- *Commitment* — These issues are directly related to the accepted TQM philosophies. They fall under the established management function of planning, as they are seen to be the base on which the whole philosophy of TQM rests.
- *Organization* — This aspect requires an organization-wide commitment to TQM. This perhaps justifies the need for an independent quality function to be established, with the authority and power effectively to influence the operation of quality-related activities.
- *Measurement* — Seen as part of the controlling function of management.

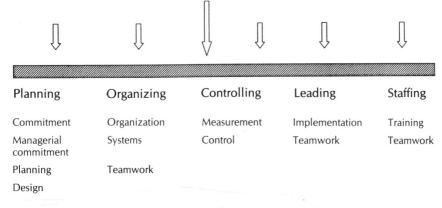

Figure 2.5 Total quality management issues.

Under the TQM umbrella, the emphasis can be directed towards planning in such a way that the costs are allocated to education, training and the prevention of mistakes (cost of quality), rather than to the errors made (cost of non-quality).

- *Planning* — In manufacturing, this is recognized as planning for just-in-time (JIT) management, purchasing and maintenance — in this text it is planning for quality in all organizational activities — including quality. In the service industries the more important aspects could possibly be flowcharting of service delivery and in particular of complaint procedures.
- *Design and systems* — These can be seen as part of the planning stage in conventional management theory. These include the design of the standards and specifications and the way these are written, together with the instruments and practices used to check the procedures. It is imperative that any management system can be used effectively. It defeats the purpose of TQM if the system itself becomes too cumbersome and bureaucratic. Thus, the development of a bureaucratic quality management system will defeat the whole purpose of TQM.
- *Control* — This covers the inspection, evaluation and rectification of systems. It is as much part of the conventional function of controlling, as it is of TQM.
- *Teamwork* — This is a specific issue in TQM. Recognized in more conventional management as one of many work environments, it is recognized by TQM as encompassing the whole organization. Without teamwork, and the development of a flexible and effective workforce, TQM will not operate.
- *Training and education* — This is a major staff issue. A systematic approach to training and education is regarded as extremely important in TQM, not only training on job knowledge, quality-related activities, the effective use of quality tools and people skills, but also to further the ability of a worker to take on more varied tasks and to work on their own personal development — a case of job enrichment. While developing more capable staff and thus being able to do a job more effectively, it motivates the staff to be more loyal to the organization, which assists them in their career.
- *Implementation* — The implementation of TQM itself must be seen as part of leadership, and most certainly other functions of quality management also.

The process of the implementation of TQM

Since organizations differ in their orientation, the application of one, 'canned' TQM system, will not generally provide the expected results of the cultural change. Using a more humanistic approach and managing in a quality way rather than managing quality, emphasis would initially be placed on statistical process control. This would invariably be implemented at every level in the institution, but initially at the production or direct service interface point. This may not occur for varying reasons, which include the effect of 'power' cultures. e.g. top management.

Although top management must drive quality, top management seems not to relish the prospect of succumbing to quality and its process trappings.

Implementation would include education in flowcharting processes and widespread statistics training, leading to the development of control limits and corrective action guidelines, by the use of systematic problem analysis to provide continuing improvement of all the processes within the organization. This will eventually lead to empowerment of individuals at the grass roots level, through the development of responsible autonomy to make their own quality-related decisions. Emphasis must be placed on understanding organizational processes and improving them, and on measurement and feedback of results in a qualitative and quantitative way.

Factors to be considered in the successful implementation of TQM in an organization suggest the unrivalled generation of the *commitment of top management* is a primary aim. The development and writing of the quality policy needs to be accomplished and communicated to everyone in the institution. There is a need to create quality visibility and educate top management in quality matters and the implications of TQM. Engendering participation and involvement from employees would seem to be a priority at this stage.

A quality council (QC) should be set up, containing all top management, especially for cross-functional quality roles and initially directing quality improvement activities. They should receive quality improvement team (QIT) reports, verbally and in writing, on a regular basis, i.e. once a month, and provide immediate direction to quality planning teams (QPTs). They should provide a means by which an exchange of ideas for continuing growth in the quality area could be determined. Once the steering group of the QC has been implemented, the set-up of QPTs is enacted. Membership should contain specialist quality and planning members and representatives from each department within the organization. The quality policy should be translated into realistic planning goals for each department and each QPT should meet regularly with the QC and the relevant department, to ensure integrity of planning direction. An important step is the development of QITs. Members for the QIT should be selected, based on problem-solving requirements of the given task, and plan the quality improvement programme. A QIT charter should be established. Each member should contribute creative ideas and effectively use problem-solving techniques, develop and use the seven tools of quality and use graphs and/or chart results using stratification methods, Pareto analysis, histograms, scatter diagrams and box plots, as required.

Another important aspect is the ability to measure quality. This is where problems and problem areas — which can include administrative or operating problems — are identified. Data about operating processes are acquired and the trends charted — a visual display of results — and effective solutions to the identified problems found. These trends are tracked to provide reinforcement for quality programme implementation. Estimating the cost of quality will ensure that management attention is given to the demonstrated process improvements.

Use can be made of the following (based on Juran, 1974), as a guide to itemizing the cost of conformance/non-conformance, when establishing a project to work on (Juran and Gryna, 1993):

Conformance costs	Non-conformance costs
Preventative maintenance	Waste
Process capability studies	Rework
Inspection	Scrap
Market survey	Administrative time-solving problems
Quality assurance	Excess delivery cost

Expansion and the creation of awareness of quality initiatives would seem to provide positive reinforcement of any process improvements, which in itself will ensure that the quality programme survives and develops. Communication of quality successes should be accomplished through regular meetings with management and employees concerning non-conformance problems. The distribution of quality-oriented posters, memos, articles, QIT results, etc., make quality results visible. There should be no start or finish; improvement must be continuous, because quality never ends. Quality awareness should be part of a planned programme and implemented by the QIT. Coupled to the improvements is the concept and practice of *corrective action*.

Systematic corrective action measures should be developed at all levels in the organization, and use made of the Pareto principle — of the vital few and the trivial many. Document corrective action taken and communicate it to other QITs in the organization.

The implementation of TQM generally means empowerment of the base workers in an organization. The vision of TQM and its application to organizations does not really provide empowerment for these people — they already have it to a large degree.

For TQM to develop, the cultural basis of the organization's structure and operating characteristics will need to change, to ensure its inherent growth and survival. Here, culture means the values, norms, behaviours, artefacts and attitudes that characterize that organization.

Flexible, multi-skilled individuals who can respond appropriately to changing circumstances and requirements are the product of the application of TQM. TQM will mean increased need for systematic planning, data generation and equitable and appropriate analysis that provide ready means for differentiating the performance of individuals. It will also mean the development of a more unified organization, where individual workers, for example, will have to adhere to group standards, rather than individual deliberation and orientations. Could this indicate a major source of resistance, where TQM may attack the individual freedoms that may be available? TQM will not of itself provide solutions to problems posed in organizations. However, TQM does offer a method whereby clear, specific and actionable measures can be introduced to increase the efficiency, effectiveness and above all the quality of work-life needed in those organizations. Whether it be TQM or continuous improvement, the goal of the organization is the same, namely achieving a standard of quality at some determined economic cost.

What is certain is the fact that, once a company embarks on the route to quality, it can never remain static, complacent or consider it has reached the summit of quality achievement. The process once started is ongoing and endless. The time period to put the basics in place can be long, perhaps ten years (Dale and Plunkett, 1991).

TQM practices — the reality

The emphasis on TQM has grown considerably over the last few years. There is now a plethora of books that have 'total quality' in their titles, e.g. *Total Quality Marketing* (Fraser-Robinson, 1991) or *Total Quality Training* (Thomas, 1992). It would seem that there is total quality in everything we do. Does this ridicule TQM, or does it actually provide further support for its magnitude of influence? The paradox is that if we believe wholeheartedly about the benefits that can accrue with the use of TQM practices, then this is a good thing. If we have reservations about the ability of TQM to deliver, then it is not.

In the 4th International Conference on TQM, Cook (1991) states that Tom Peters said,'Most TQM programmes fail because they have system without passion or passion without system'. It means that introducing the quality management system, without thinking about its consequences on people, will lead to failure.

It is interesting to note that it has been demonstrated over the past fifty years that it is easier to control systems than it is to control people. Occasionally, management ensures failure by doing just this; they concentrate on systems without equally concentrating on people. One of the elements of TQM is teamwork, and this should involve people in the concept of TQM right from the start. Large and small companies have implemented TQM with varying degrees of success. It would seem that larger organizations have a better track record than smaller ones, but possibly because of necessity — competition — and because of forward-looking orientations, coupled with the required resourcing capabilities.

Since larger organizations have more money available for developing TQM, have more to gain with its implementation, have a longer strategic outlook, and can afford the continuous costs afforded to nurturing TQM practices, it is natural to appreciate that these larger organizations experience more positive TQM realizations than smaller ones.

Sector experiences

- MANUFACTURING

Manufacturing-focused organizations have been at the forefront of managing quality throughout this and the last century. The development of TQM has been on the back of what was learned from managing manufacturing processes. Even so, for many organizations managing in a quality way is a first attempt at the implementation of TQM. Manufacturers produce product — something that can be touched. Consequently, the focus for these types of organizations has always been the production output.

The quality-oriented manufacturer must learn to change this focus. It means changing first to a customer, then to an internal process orientation to ensure that customers get what they want. This development has been led by possibly three factors — customers, technology and competition. However, much change in quality practices has been targeted towards what technology offers. It is only of late through heightened competition, that the customer has been considered.

What are the problems of implementing TQM in a manufacturing organization? Many manufacturing organizations are adept in the application of quality control and quality assurance. This is based on their fundamental focus, the manufacturing process. The Japanese have excelled in this type of application. TQM requires organizations to take a major step forward; that is, to nurture their staff, become flexible and let lower staff make more meaningful decisions. This is the problem. The culture of a standardized organization — in respect of its manufacturing focus — is very difficult to change. It is not sufficient to produce product to designed specifications, when those specifications do not meet the customer's needs.

TQM, therefore forces a manufacturer to become externally oriented, as well as strategically oriented, as productivity problems (quality problems) are related to the objectives of what must be achieved. Integration is the name of the game; cross-functional teams are necessary, rather than a group of isolated experts.

The problems seem to indicate exclusive attention to:

1 Internal — product — focus.
2 Standardized practices, that are inflexible and lead to a process-oriented culture.
3 Produce first, correct problems later. This results in increased external failure costs.
4 Marketing seen as a selling activity.
5 Internal-orientation, such that we do the best we can in the circumstances.
6 We know what our customers want, and we manufacture product to meet those requirements.

So what can TQM do for a manufacturer? TQM offers manufacturing the ability to:

1 Meet, exceed and eventually to anticipate customer needs.
2 Develop a whole organization approach to managing quality and to break the barriers between production and other traditional service elements in the organization.
3 Change the application of manpower engaged in end-of-line inspection of

product (quality control) to on-line inspection applied at time of manufacture.
4   Develop more proactive techniques in order to manage more effectively processes related to marketing, engineering design and manufacturing, in order to measure and communicate exactly what the customer wants and deliver it.
5   Ensure that marketing is an integral activity that has the first responsibility of ensuring communication with the customer.
6   Make decisions that are made of real data, derived from all business processes.
7   Seek actively and use benchmarking of internal processes, compared with other internal processes, and also with external processes occupied by industry leaders.
8   Provide education and training to staff beyond the immediate job limits and scope of workers and staff — especially quality-related techniques such as statistical process control (SPC).

- SERVICES — EDUCATION

One of the elements influencing TQM is statistical process control (SPC). When applying this to educational institutions, it would seem to be prudent to relate the kinds of metrics traditionally used to the quality requirements needed under TQM.

The general experience of educators with these kinds of metrics would make it easier for say, lecturers, to acquire the skills and capabilities to solve problems and seek improvements. As with simple education principles — taking students from the known to the unknown — the same would apply with the implementation of SPC with lecturers. Basic improvements in the way an educational institution operates require a different approach to the meaning of quality in education.

Previously, quality in education has meant quality of conformance (similar to manufacturing) to design standards — curriculum standards — and the metrics developed has reflected this. Therefore, quality assurance was the expected norm. With TQM implementation, this practice is insufficient and thus forces the development of quality initiatives that reflect this requirement. Quality of design should take greater importance than quality of conformance of delivery of a course.

In the 2nd International Conference on TQM, Curtis (1989) discusses the implementation of TQM practices at Avon Industries. For Curtis, TQM means the dependence on the partnership between management responsibility and employee commitment. Thus for education, it means that no matter what the strategic direction in which top management want to take the institution, it cannot be accomplished by themselves. The partnership includes all employees, not just those committed to quality, right through the organization. Quality improvements must be effectively managed and reinforced. This, coupled with the general acceptance that some lecturers are carrying out cutting-edge research activities, means that the generation of commitment is paramount.

The CEC draft publication (1992) states that 'There are problems in implementing TQM within an educational institution because the criteria for quality in a company do not necessarily apply in every detail. Education has its own set of values and practices, and has a different focus and objectives'. The development of an educational quality model has to reflect these differences.

Saunders and Walker (1991) suggest that individuals in educational institutions are 'more likely to question and debate the philosophy of TQM and its challenge to accepted management theory'. This may be true in most other organizations. It would seem that the implementation of TQM may actually create problems of bureaucracy that can be seen as detrimental to the 'effective' operation of each working team or group.

Why is quality performance so important to educational institutions? It should be easy for educational institutions to introduce and effect the change required by TQM. The rationale for this is that educational institutions are, by their very business nature and volition, in the business of education and change.

Since TQM requires a high investment in education and training, what better organization to do it? If it was this easy, all educational institutions would have successfully implemented TQM programmes, but they have not, first because these organizations have not been managed in a total quality way. They therefore, have to learn *how to learn* about quality and its implications, just as any other organization, in order to survive and grow. Second, how can performance in an educational institution be effectively evaluated? What seem to be fair evaluation practices in one part of an institution may not be fair in another. Consequently, quality issues need also to focus on internal machinations and politics as well as determining effective quality measures that can empirically demonstrate improvement in process deliberations.

2.3 Chapter review

Quality management has developed over the past 100 years or so. The first era of quality management was characterized by the application of quality inspection techniques. Scientific management provided the backdrop for its necessary development. Mass production brought pressure to produce consistent quality goods and this was translated into an effective strategy by inspecting quality into the process output. Quality was also introduced into each product by ensuring that workers carried out simple tasks so that errors were minimized. This was achieved through breaking down each job into its simplest task. Consequently, there was a power shift to end-of-line inspection staff. As befits the scientific management approach, even inspector's jobs were simplified and thus the limits of inspection were discovered.

The second era of quality management was characterized by controlling the manufacturing process through managing data. During this era, Shewhart developed the quality control chart in order to make it easier to achieve this. Shewhart applied the principles and practices of probability to manufacturing processes, as he recognized that variability was inherent in a process and to manage that process effectively required companies to manage the variability. The major difference between inspection and control is focus — product in the first era and process in the second. Sampling techniques were also developed in this era, but their application was limited to end-of-line operations.

The third era of quality management was characterized by developing the system that surrounded the process and the manufactured product, in which a broader approach was gained. Specialists were no longer capable of managing quality on their own and this provided the impetus and the necessity to increase its application across the organization. Quality systems, such as BS EN ISO 9000, were developed and applied.

The fourth era of quality management was the development of total quality management. TQM is a philosophy that seeks to gain organization-wide commitment, through participation, to manage quality effectively so that errors are minimized and customers are consistently satisfied. TQM is seen to require the balancing of five systems — process, technology, people, task and structure.

The problems raised by attempting to implement TQM are considerable and need to be addressed early in the quality change programme. Implementation issues include consideration of the organizational structure, management commitment, management and worker culture. Implementation elements:

1. Generate the commitment of top management and develop the quality vision and the necessary quality leaders.
2. Form a quality council, planning teams and quality improvement teams.
3. Collect data and estimate the cost of quality.
4. Develop a quality culture through effective problem-solving techniques and corrective action programmes.

The implementation of TQM essentially means the empowerment of base workers through increased communication, education and training.

2.4 Chapter questions

1. Define the term TQM and outline the benefits to an organization implementing the management philosophy.
2. Outline the basis and development of TQM.
3. Discuss the various issues surrounding the implementation of TQM.
4. Describe a model of the process of implementation of TQM.
5. Outline and evaluate the necessary steps towards the implementation of TQM.

CHAPTER 3

Quality management writers

CHAPTER OBJECTIVES

Discuss the content and impact of various quality management writers

CHAPTER OUTLINE

- **3.1** Quality management writers
  - 3.1.1 *Juran*
  - 3.1.2 *Deming*
  - 3.1.3 *Garvin*
  - 3.1.4 *Crosby*
  - 3.1.5 *Ishikawa*
  - 3.1.6 *Feigenbaum*
  - 3.1.7 *Taguchi*
- **3.2** Chapter review
- **3.3** Chapter questions

3.1 Quality management writers

Every management discipline has its writers. We have encountered many in the first chapter dealing with the development of management. In this chapter, we will discuss briefly a number of individuals who have become known as influential quality practitioners. The discussion here will be limited to individuals who have demonstrated many years of commitment to quality in many forms and many ways. Quality management practitioners, their major philosophies and orientation can be seen in Table 3.1.

Table 3.1 Quality management practitioners' orientation

Writer	Definition of quality	Orientation	Developed
Juran	Fitness for use	Customer	Quality trilogy The five quality characteristics Internal customer The four phases of problem solving Quality council The quality spiral
Deming	Fitness for purpose	Customer	Fourteen points of quality Deming PDCA cycle Seven deadly diseases System of profound knowledge
Garvin	None specific	Customer and supplier	The five bases of quality Eight dimensions of quality
Crosby	Conformance to requirements	Supplier	Five absolutes of quality Fourteen-point plan for quality
Ishikawa	None specific	Supplier	Fishbone diagram Classification of statistical quality tools Company-wide quality control Quality circles
Feigenbaum	Customer satisfaction at the lowest cost	Supplier	Industrial cycle Utilization of the quality consultant
Taguchi	None specific	Supplier	Quality of design methods

3.1.1 *Juran*

Juran's definition of quality is *fitness for use*. He uses this in the context if a user-based view signifies that quality lies with the actual use of a product or service. Juran applied two different meanings to quality — features and freedom from deficiencies. Effectively managing these quality types means using what now seems an age-old concept of his quality trilogy (Juran, 1986). This concept can be seen in Figure 3.1, and indicates the connection between quality planning, quality control and quality improvement.

Only the customer can determine the quality of the product or service when using this definition. Consequently, manufacturers do not like to use it, but prefer a more controlled conformance to specifications. Therefore fitness for use is a utility value concept which varies from one customer to another.

According to Juran (1974), this concept was based on the following five quality characteristics:

1   Technological (e.g. strength).
2   Psychological (e.g. beauty).

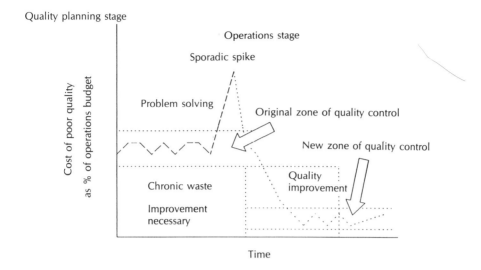

Figure 3.1 Juran's quality trilogy. Reproduced with permission of the copyright holder, Juran Institute, Inc., 11 River Road, Wilton, CT 06897 USA. All rights reserved.

3   Time-oriented (e.g. reliability).
4   Contractual (e.g. guarantees).
5   Ethical (e.g. sales staff courtesy).

The quality for a manufactured product may be defined primarily by technological and time-oriented characteristics, whilst a service product may involve all the characteristics listed above. This is an example of why it has been difficult to implement quality programmes in service industries. Further, Juran determined that fitness for use can be broken down into four elements — quality of design, quality control, availability and field service — as shown in Figure 3.2.

Juran also derived the concept of the internal customer, which relates to an organization with more than one person. Internal customers were individuals that were supplied by downstream processes. This means that the concept could be applied to physical product or just information flow. Each upstream customer had specifications that needed to be met by downstream suppliers and all these internal customers were working towards external customer satisfaction. Process analysis would therefore help to satisfy external customers by making the internal organization more effective.

Morgan's (1986) machine metaphor would not look out of place in this concept, since, when applied to an organization, it would mean that individuals tend to act out three roles — customer, processor and further supplier. Oakland (1989) called this the *internal customer chain*.

Juran's focus was always quality improvement. Here, he determined that the goal was to increase performance to levels never achieved previously. Juran

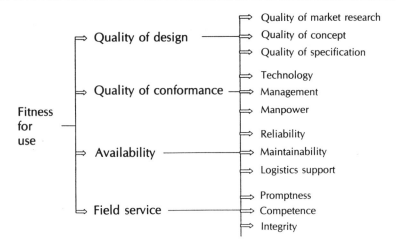

Figure 3.2 The four elements of Juran's fitness for use.

suggested that in order to do this — a project — by working on a problem, companies must go through a series of breakthroughs in attitude, organization, knowledge, cultural patterns and results (Juran, 1964). Consequently, he developed the six phases of problem solving for quality improvement. These were broken down, as seen in Table 3.2.

Phases 1, 2 and 3 could be thought of as the journey from symptom to remedy. Phases 4, 5 and 6 could be thought of as the journey from remedy to further opportunity. These are further illustrated in Figure 3.3. This process was cyclic in nature and reflected the continuous spiral of quality development in an organization.

Juran was the first to point out that the Pareto principle could be applied to quality improvement. The basis is to distinguish the important *vital few* from the *trivial many*. One of the tools of quality used by Juran is the Pareto principle. Juran was also very interested in the cost of quality, and the Pareto tool was used extensively to illustrate to top management the effects of improving (in cost terms) the vital few. Juran also introduced the development of the quality council (see Chapter 2), a body that manages the quality activities of an organization. It is here that all quality activities are sanctioned and directed.

3.1.2 *Deming*

Deming is remembered for his fourteen points, the Deming cycle, and his deadly diseases. Earlier than many, he had a particular appreciation of statistics. In the 1950s Deming taught the Japanese statistical process control. In recognition for his timely intervention and contribution to Japanese industry, the Union of Japanese Science and Engineering instituted the Deming prize. In 1980, the Metropolitan section of the American Society for Quality Control established the

Table 3.2 Six steps to problem solving. Reproduced with permission of the copyright holder, Juran Institute, Inc., 11 River Road, Wilton, CT 06897 USA. All rights reserved.

Step	Activity
1. Identify the project	• Nominate projects • Evaluate projects • Select a project • Ask: 'Is it quality improvement?'
2. Establish the project	• Prepare a mission statement • Select a team • Verify the mission
3. Diagnose the cause	• Analyze symptoms • Confirm/modify mission • Formulate theories • Test theories • Identify root cause(s)
4. Remedy the cause	• Identify alternatives • Design remedy • Design controls • Design for culture • Prove effectiveness • Implement
5. Hold the gains	• Design effective controls • Foolproof the remedy • Audit the controls
6. Replicate and nominate	• Replicate the results • Nominate the new projects

Deming Medal to be awarded for achievement in statistical techniques for improvement in quality.

Deming's fourteen points are:

1 *Create constancy (and consistency) of purpose.* This means management is charged with the requirement to plan for today and tomorrow and to provide a co-ordinated and organized effort to reach the quality goals set for tomorrow. It also means getting rid of short-termism and use quality planning effectively. The long-term plan, when used as a guise to generating short-run paybacks, is begging the organization to trip and fail — efforts will be short-sighted and gains will be lost. The quality-oriented organization will undoubtedly want to secure its future and will respect the fact that quality development is a longer term process. Deming suggests that constancy of purpose means innovation, research and education, continuous improvement of product and service, and maintenance of equipment and plant (Walton, 1986).

2 *Adopt the new philosophy.* The quality culture must become part of the

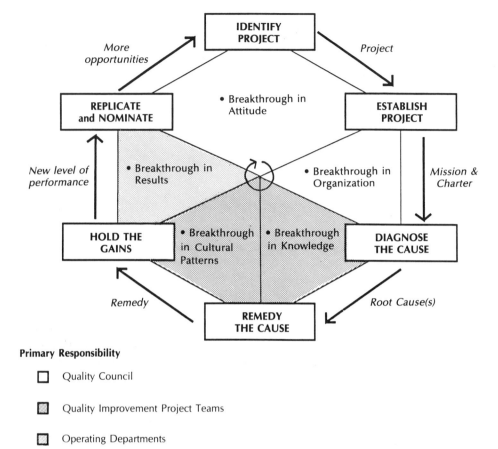

Figure 3.3 The six major steps of problem solving. Reproduced with permission of the copyright holder, Juran Institute, Inc., 11 River Road, Wilton, CT 06897 USA. All rights reserved.

patchwork of the organization. It must become a sort of religion that espouses an excitement that makes everyone in the organization happy to be part of. Deming suggests that this is essentially a transformation of management. Since quality will transform workers into self-managed units, it is about management at all levels — management and the workforce. The demarcation line between management and workers will decrease as a consequence.

3. *Cease dependence on mass inspection.* In its era, mass inspection was all the quality management we had. This is no longer the case. Inspection costs. It costs, not only in the wages of the inspection team, but also in regard to non-prevention. Inspection does nothing to prevent the occurrence of

the defect again and again. Deming once said 'that quality comes not from inspection but from improving the process'.

4   *End the practice of awarding business on price only.* This is a major problem that supports short-termism. The practice generates many more suppliers than necessary and the relationships between the organization and the suppliers are strained as a result. Managing the constant flow of product that meets specifications through inspection becomes a problem, and inevitably leads to defective product moving through the production system.

5   *Improve constantly the system of production and service.* This is similar to all quality management writers — constant process improvement. Deming (1986) said that 'quality must be built in at the design stage'. It means moving away from the status quo — and that means risk; risk for management and risk for workers, risk that needs to be made to ensure continual and effective quality development. The use of the PDCA cycle would be extremely useful in this context.

6   *Institute training and retraining.* Train and educate in the principles and practices of quality management, including SPC and the tools of quality for job-specific requirements. This gives them the confidence to experiment under the knowledge of the processes they control. Systematic training needs to be an integrated part of the work ethic, which encompasses the best training practices and sufficient resources allocated to ensure effective skill development.

7   *Institute leadership.* The generation of effective leadership, especially self-leadership, is a primary requirement in the quality culture. This will become more important as work groups are set up in the organization to work on quality problems — seemingly independent of top management.

8   *Drive out fear.* The blaming culture needs to give way to a problem-solving culture. This means that suggestions for improvement must be seen in a positive light and that responses from management, short of full support, must be avoided. Established bureaucracies seem to be the worst offenders, who seem to discriminate against 'free thinkers' and other people who would like to do something positive about their own work situation, but are shackled by peers and managements' own fear of consequences.

9   *Break down barriers between staff areas.* Structural barriers in an organization should be removed if these interfere with effective quality improvement. Particular areas include the design stage of a new product, where engineering, marketing and production vie for control of the new product. Barriers prevent ownership transference and reflect bureaucratic 'turf' keeping.

10 *Eliminate slogans, exhortations and targets.* Slogans reflecting ideal

situations that bear no resemblance to the present quality situation should be avoided. Deming (1986) suggests that slogans have a lofty ring — 'zero defects' — for example, 'but how could a person make it right the first time, when the incoming material is off-gauge, off-colour or otherwise defective'? Similarly, numerical targets need to have the support of good process and good equipment and training. Setting unrealistic targets will demotivate staff not motivate them to work harder.

11 *Eliminate numerical quotas.* Deming (1982) said 'I have yet to see a work standard that includes any trace of a system which would help anyone do a better job'. Applying work standards have demonstrated much frustration for the workforce, as effort is not included in the output criteria. This demotivates staff and sets up barriers to improvement. Natural quotas will develop as a result of the limiting capability of the process and the equipment, not as the limiting motivation of the employee.

12 *Remove barriers to pride of workmanship.* Managers pushing numbers of output, rather than the quality produced is a major barrier to pride development. Individual incentive schemes, conflict and misunderstanding between departments, lack of training, standards changing all too frequently and supervisors, all give rise to barriers to pride in their work.

13 *Institute a programme of education and retraining.* Continuous improvements in quality will lead the workforce to develop new skills, new ways of carrying out their job. As these skills develop, management must develop with them and provide education and retraining — not as an expense but as an investment in their own and the organization's future. As the jobs develop, a newer, more rationalized workplace may result. People will need to be trained and educated to fit into these new jobs effectively. Education is a way of improving people as well as improving their capability of carrying out their workplace roles.

14 *Take action to accomplish the transformation.* This requires top management commitment and a long-term orientation. It will not happen today or tomorrow! The initiative for engaging the other thirteen points must come from top management. Deming (1986) suggests that the use of PDCA is a universal means to quality improvement. Initially called the Shewhart cycle, this has become known as the Deming cycle, after it was shown in the 1950s to many Japanese managers and engineers.

Deming suggests that American (Western) management suffers from a number of deadly diseases that conspire to prevent effective management practices being developed. These were:

1 *Lack of constancy of purpose.* This reflects short-termism and is seen by Deming as a fault of management being guided too closely by accountants

and financiers. Also inconsistency with quality policy and plan implementation.

2   *Emphasis on short-term profits.* This is related to number one above. Short-termism is also about extraction of funds that could be reinvested in the people that matter in an organization — its workforce. Public companies are likely to have to balance shareholder profits, takeover bids, the workforce and the customer — and the shareholder generally wins, to the detriment of the whole organization and ultimately the customer.

3   *Evaluation of performance or annual review.* The use of management techniques such as management by objectives causes much short-term planning and fear when negotiating job performance requirements. They create conflict between team members and assist mediocre group performances, rather than performances that reflect the capabilities of each individual. It means people individually moving to targets they agreed individually. Competition is rife. The bad negotiator does not mean a bad worker, but because of the power relationships of supervisor and worker, the worker is penalized.

   Deming obviously does not like evaluations, to the extent that all his students would receive grade As for the term work they handed in.

4   *Mobility of top management.* Deming thought that too many top managers were in 'temporary' posts of two to three years — hardly any time to evaluate their credibility in managing the organization effectively. Mobility may be okay for the manager, but it does not provide consistency of the top management rhetoric. Movement of top managers and workers is seen as a reflection of underlying problems of dissatisfaction with the job. Managers do not seem to give themselves enough time to learn the 'real' problems of the job before moving on.

5   *Running a company on figures alone.* The measurement of the objective side of an organization is a relatively simple task these days. It is the unknown figures that create the problems, e.g. how can you cost the effect of a dissatisfied customer on your sales potential?

6 and 7   Excessive medical costs and excessive costs of warranty fuelled by lawyers. These are taken as beyond the scope of this discussion.

Some of the obstacles Deming suggests do not help to develop effective quality management include:

1   Our problems are different.
2   Reliance on quality control departments.
3   Quality by inspection.
4   Blaming the workforce.
5   Inadequate testing of prototypes.

Deming has attempted to review his quality management philosophy and describes this as:

1   *Appreciation for a system* — This means everyone needs to understand the constituent parts of the system in which they work and the various interrelated relationships that occur; one failure in one part of the system affects success in another part.
2   *Knowledge of statistical theory* — Deming requires that all staff are conversant with the general methods of statistics and able to apply them effectively.
3   *Theory of knowledge* — This relates to effective planning and implementation of those plans to determine what works and what does not.
4   *Knowledge of psychology* — Quality development requires changes in people's attitudes, values and behaviours. Consequently, management and workers alike need to understand what drives people and how these drives can be tapped for the continuous development of quality.

3.1.3 *Garvin*

Garvin is a Professor at Harvard Business School. He has developed a number of contributions that greatly influence quality management theory. He developed what has become known as the eight dimensions of quality. These dimensions are performance, features, reliability, conformance, durability, serviceability, aesthetics and perceived quality. They are understood to indicate the breadth that quality has come to mean, and in this context suggest that multi-dimensions are required to elicit even the most fundamental meaning of quality.

Garvin (1988) also introduced the notion of the five quality bases — transcendent, product, user, manufacturing and value. These major contributions are further discussed in Chapter 4.

3.1.4 *Crosby*

Crosby is perhaps the quality equivalent of Tom Peters to general management. He has a flair and style that is synonymous with a get-up-and-go mentality. In effect his energy as a prolific writer and facilitator means that the quality message is worth listening to.

His book *Quality is Free* depicts a viewpoint that the cost of running a quality programme in an organization can be more than offset with the financial gains of satisfied customers. Crosby's quality slogan is 'conformance to the requirements and quality is free'.

Crosby (1979) developed what he called the five absolutes of quality. These are:

1   Conformance to requirements. The idea behind this is that, once the requirements have been determined, the production process will exhibit

quality if the product or service resulting from that process conforms to those requirements.
2   There is no such thing as a quality problem.
3   There is no such thing as the economics of quality; it is always cheaper to do the job right first time.
4   The only performance measurement is the cost of quality.
5   The only performance standard is zero defects.

The underlying philosophy behind these absolutes is a conformance mentality. It breaks down if the design of the product or service is incorrect or does not match the actual customer requirements effectively. Since management deals predominantly in the language of money, putting the cost of non-conformance in these terms makes sense. It clearly illustrates the effect of non-conformance and focuses attention on prevention issues. This is Crosby's basic thesis behind quality is free.

Crosby's fourteen-point plan for quality improvement deals predominantly with implementation issues. Deming's fourteen-point plan rests squarely with a driving management philosophy.

The fourteen steps of Crosby are:

1   *Management commitment* — Determining where management stands on quality, developing a quality policy and management visibly becoming serious about quality.

2   *The quality improvement team* — Crosby suggests that all members, except the chairperson, are part-time — because of the time commitment. It could be said that he also means that the QIT, like the rest of the organization, is mentally attuned to quality 24 hours a day. He also indicates the responsibilities of the team members, which include:
    (a) Develop and action the quality improvement programme.
    (b) Represent their department fully on the team.
    (c) Co-ordinate and execute quality decisions made by the team that affect their department.
    (d) Contribute creatively to the quality programme.

3   *Quality measurement* — According to Crosby, this means generating data about current and potential non-conformities and developing appropriate corrective actions. Measurement data must be current and preferably on line. The use of the most up-to-date information will ensure effective quality decisions. To overcome waste in the various processes — manufacturing or service — Crosby suggests three things — recognition of the problems, measurement of current status and a developed quality programme for reducing the waste.

4   *The cost of quality* — Crosby indicates that the cost of quality includes scrap, rework, warranty, inspection and quality control labour, design and/or

engineering changes, etc. and audits. Crosby suggests that cost of quality is a 'catalyst that brings the quality improvement team to a full awareness of what is happening'.

5. *Quality awareness* — This means providing the sort of support necessary to raise the level of concern and interest in quality amongst all staff in order for them to understand, acknowledge and support the reasons for the quality programme. Crosby indicates that awareness needs to be raised for the conformance to requirements of the processes to be accepted and to prepare them for zero defects programme. The awareness programme consists of two major activities — regular quality-oriented meetings between management and employees and the communication of information about the progress and extent of the quality programme and related initiatives. Quality awareness should tend towards being low-key, but invariably with constant attention.

6. *Corrective action* — Crosby states that there is a need to develop systematic methods to solve problems previously exposed. This he suggests should be carried out through four levels of constant activity — daily, weekly and monthly meetings and tasks teams that work and meet daily until the identified problem is solved.

7. *Zero defects (ZD) planning* — Crosby (1979) indicates that the main points of ZD planning are:
   (a) Explaining the concept and programme to all supervisors.
   (b) Determining what material is required.
   (c) Determining the method and process of delivery of the ZD programme.
   (d) Identifying the error–cause–removal programme and making plans for its execution.

8. *Supervisor training* — Crosby suggests this is necessary in order to ensure that supervisors are able to carry out the tasks and responsibilities of the quality improvement programme. Crosby indicates that supervisor training is divided into three parts:
   (a) Supervisor training covering quality measurement techniques, costs of quality implications, corrective action methods and the quality awareness action.
   (b) Zero defects programme briefing.
   (c) Do it over again.

9. *ZD day* — This means making a given day a visible connection between the quality rhetoric and promises of the past few months and the future commitment and understanding from all concerned — from that point onwards. The essential point is that management have publicly committed themselves to quality and are expecting the workforce to do likewise.

10. *Goal setting* — A requirement that creates motivation and the drive to

11 *Error cause removal* — This is a systematic method of ensuring that the employee can communicate to management the quality problems that affect them carrying out their job. Crosby notes that *every* single response should be taken seriously. These can range from very simple to very complex problems — both types need the commitment of management.

12 *Recognition* — According to Crosby, people do not just work for money. Consequently, Crosby determines that recognition in other forms is seen to be more appropriate in a quality environment.

13 *Quality councils* — Crosby says that this is to 'bring together the professional quality people for planned communication on a regular basis'. This is fine if this refers to all people working for quality as quality professionals — but if it refers only to the individuals who are trained professionally, then this is misleading. The quality council is a very important part of the quality development in an organization. It must contain a balance of professionals and people working at the heart of quality — the shop-floor.

14 *Do it over again* — Emphasizing that quality is about continuous improvement.

These steps are not actually steps in the sense that you move from one to the other. They should be used as a guide to help in the development of a quality programme.

Crosby and Deming have a similar message in regard to top management. It is they who must take the blame for poor quality procedures and outcomes, and it is they that can make sure that no poor quality exists in the organization. Crosby's approach is therefore top-down management of quality with education in quality for all staff, irrespective of their position in the company.

3.1.5 *Ishikawa*

Ishikawa was probably best known for his contributions to quality management through statistical quality control. His development of the Ishikawa diagram (fishbone) and the employment of the seven old tools of quality provided grass-roots capability in the use of problem-solving techniques.

Ishikawa developed a simple classification of statistical quality tools, which were hierarchical in nature, in the sense of the statistical expertise required to apply them. This was:

1 In the lowest — seven tools — these are tools capable of being learnt and applied by everyone in the organization. This means that shop-floor

personnel would have the statistical capability to evaluate quality problems. These tools include:
(a) Cause and effect diagram.
(b) Pareto analysis.
(c) Stratification.
(d) Histograms.
(e) Process control charts.
(f) Scatter diagrams.
(g) Check sheets.

2 The next are the tools that can be used by managers and quality specialists — they include hypothesis testing, sampling, etc.

3 The last grouping could only be used for advanced statistical problem solving for use by quality specialists and consultants — they include experimental designs (Taguchi methods) and operations research techniques. These are highly mathematical in nature and few people have the necessary background to apply them effectively, hence their limited use in organizations.

Ishikawa was more people-oriented than statistically oriented. His main aim was to involve everyone in quality development, not just the management who drove it. The heart of his contributions was the attention he gave to problem solving. Consequently, the importance given to the tools of quality cannot be overemphasized. Nevertheless, Ishikawa's reliance on generating data about processes and the use of simple statistical techniques illustrates his down-to-earth methods.

Ishikawa believed that a Western lack of attention to the contribution everyone can make to quality in an organization had made Western management thinking in quality rather less of an impact to what it should have been. It means that grass-roots workers were denied, and are still denied in many organizations, a contribution to quality. The Japanese insistence on teamwork, and all staff being 'equal' on the basis of contributions to quality, illustrates the major gap existing between Japanese and Western management quality practices.

Ishikawa insisted on the idea that customers' complaints were opportunities that should never be wasted. They were an opportunity for quality readjustment and that seeking customer complaints, although controversial at the time (twenty years ago), must be encouraged in order for the organization to develop in the direction it should — satisfying customers and perhaps even delighting them.

There was also an organizational orientation to Ishikawa's contributions in the form of *Company-wide Quality Control* — following Juran and Deming's visits to Japan in the 1950s. This further emphasized Ishikawa's thinking, where all in an organization were trained in statistical techniques, from top management to the shop floor. Company-wide participation indicated that management was committed to quality management and that everyone's contribution to solving problems — however small or large — was significant.

Ishikawa has perhaps become well known as the *father* of quality circles — that much despised (in the West) group of individuals organized into relevant work teams that solve quality-related problems. These are the major groups who use the seven old tools of quality. Quality circles developed as a consequence of the implementation of company-wide quality control measures. In effect, they were smaller work units where everyone could do everybody else's job and therefore multi-skilling was enhanced.

3.1.6 *Feigenbaum*

An engineer, Feigenbaum (1991) became known for his work in quality control. As early as the 1950s he defined total quality as 'an effective system for integrating the quality development, quality maintenance and quality improvement efforts of the various groups in an organisation so as to enable production and service at the most economic levels which allow customer satisfaction'.

Feigenbaum originated the industrial cycle — the development of a product from concept to market launch and beyond. This cycle included marketing, design, production, installation and service elements, now considered essential elements in the management of quality in an organization, as well as in managing a quality management system such as BS EN ISO 9000.

His total quality view did not extend to the application of quality improvement responsibility to all employees in the organization. This ideology conflicts sharply with the philosophies behind TQM. Nevertheless, the quality viewpoint is a major contribution to quality management thinking. Central to the issue of managing quality is the use of the quality professional as co-ordinators and supporters of the total quality management process. The total quality view was based on the notion of total cost and that managing in a total quality way would result in lowered overall cost to the organization and therefore to the customer.

Feigenbaum also introduced the concept of the *hidden plant*. This introduced the idea that waste lowered the real capacity of a plant because of rework and not actually getting it right first time. Today, figures in the region of 20 per cent mean that customers could get product and services for up to 20 per cent less — effectively ensuring greater market share and an increased bottom line.

3.1.7 *Taguchi*

Taguchi's major contribution is about the effective quality of design. Essentially, Taguchi's methods focus on determining the *cost* of not meeting the specified targeted value. This conflicts with traditional quality management practices, in that, as long as a product meets the specification limits set for a given product/process (tolerance), then the product is acceptable. Taguchi could not agree with this viewpoint. Consequently, he developed the *loss function*, where he calculates reducing utility as a function of the distance from the target value a product or process characteristic results — which is the loss to society in terms of cost. The loss function approximates to the square of the distance from the target value.

Taguchi believes in designing a product and the production process to achieve a target value by making it robust and insensitive to process variations. In order to implement his idea in this area, Taguchi uses parameter design and control of experiments techniques. Here, Taguchi positively favours the more proactive quality practice of off-line quality control through effective design and development.

He suggests that the time and effort spent designing and planning will save much more effort, time and cost later, during on-line quality control. The most effective end design product he suggests, results from the use of three stages. Stage 1 — system design, stage 2 — parameter and robust design, and stage 3 — tolerance design.

Taguchi's developments are considered further in Chapter 13.

3.2 Chapter review

Every management discipline has its writers and quality management is developing into a very important specialist area. Many important contributions have been made over the past 100 years or so and it is a difficult task to separate the many contributors and develop simple monographs of their theories and applications.

Juran supported the definition of quality as fitness for use, which can be subdivided into quality of design, quality of conformance, availability and field service. The concept is based on five characteristics, which are technological, psychological, time-oriented, contractual and ethical. Juran also introduced the concept of the internal customer. Juran's focus was always quality improvement, which was broken into project definition, the diagnostic journey, the remedial journey and holding the gains. Further, Juran introduced the notion of the quality council.

Deming introduced the fourteen points, the Deming cycle and the seven deadly diseases. He also introduced his system of profound knowledge. Deming supports the definition of quality as fitness for purpose.

Crosby supported the definition of quality as conformance to requirements. He also introduced the notion that quality is free.

Crosby developed what he called the five absolutes of quality, in which the underlying philosophy behind these is a conformance mentality. He also developed a fourteen-point plan for quality.

Ishikawa was probably best known for his contribution to quality management through statistical quality control. The development of what is now known as the Ishikawa (fishbone) diagram and the employment of the seven old tools of quality provided grass-roots capability in the use of problem-solving techniques. Ishikawa's main aim was to involve everyone in quality development, not just the management which drove it.

Feigenbaum became known for his work in quality control and originated the industrial cycle — the development of a product from concept to market launch and beyond. However, his total quality view did not extend to the application of quality improvement responsibility to all employees in the organization, limiting the extent to which his theory could be applied today. Although his view suggests that the use of the quality professional as co-ordinators and supporters of the total quality management process is central to the issue of managing quality — a technique much used today. Feigenbaum also introduced the concept of the hidden plant, the idea that waste lowered the real capacity of a plant because of rework and not actually getting it right first time.

Taguchi's major contribution is about the effective quality of design. Essentially, his methods focus on determining the cost of not meeting the specified targeted value. This conflicts with traditional quality management practices, in that as long as a product meets the specification limits set for a given product/process (tolerance), then the product is acceptable. Consequently, he developed the loss function, where he calculates reducing utility as a function of the distance from the target value a product or process characteristic results, which is the loss to society in terms of cost. The loss function approximates to the square of the distance from the target value. The most effective end design product he suggests, results from the use of three stages. Stage 1 — system design, stage 2 — parameter and robust design, and stage 3 — tolerance design.

3.3 Chapter questions

1. Compare and contrast Juran's *fitness for use* with Deming's *fitness for purpose* and Crosby's *conformance to requirements*.
2. Explain Ishikawa's fishbone diagram.

3 Discuss Juran's six phases of quality improvement.
4 Discuss Deming's seven deadly diseases.
5 Contrast Deming's fourteen points with Crosby's fourteen steps of quality.
6 Outline the accomplishments of Feigenbaum and Taguchi. Determine how they may affect quality management practices in a service organization.

CHAPTER 4

The three views of quality

CHAPTER OBJECTIVES

Evaluate the three quality views
Compare and contrast the five quality bases of quality
Discuss the relative position of the customer in relation to these quality bases
Evaluate the effect of the differing views of quality
Discuss the factors that affect customer perception of quality
Compare and contrast the five bases of quality with the eight dimensions of quality

CHAPTER OUTLINE

4.1 Introduction
4.2 The three quality views
4.3 The five quality bases
    4.3.1 *Transcendent quality view*
    4.3.2 *Product-based quality view*
    4.3.3 *User-based quality view*
    4.3.4 *Manufacturing-based quality view*
    4.3.5 *Value-based quality view*
4.4 Where does the customer fit into all this?
4.5 The effect of the differing views of quality
4.6 Factors affecting customer perceptions of quality
    4.6.1 *Performance*
    4.6.2 *Features*
    4.6.3 *Reliability*
    4.6.4 *Conformance*
    4.6.5 *Durability*
    4.6.6 *Serviceability*
    4.6.7 *Aesthetics*
    4.6.8 *Perceived quality*

4.7   A comparison of the five quality bases and the eight dimensions
4.8   Chapter review
4.9   Chapter questions

4.1 Introduction

Work practices have developed from a craft-based strategy — in preindustrialization — through to multi-skilling and general management requirements on the shop floor. It would seem that we have come almost full circle. Once again, craftsmen (craft workers) are taking their place and designing and producing goods and services. But what does quality mean? To you? To me?

This chapter outlines the various quality bases that have been derived over the years. This provides some understanding as to the scope of the quality base, but does not indicate the importance of each or any of these bases, except to say that manufacturing companies have predominantly pushed their quality base as the most important, as we shall see.

4.2 The three quality views

These are the:

1   Psychologically related quality view — transcendent-, user- and value-based quality views.
2   Process-based quality view — manufacturing or service.
3   Product- or service-based quality view.

These are illustrated in Figure 4.1.

The underlying basis for these three quality views is Garvin's five quality bases, and it is here that the discussion is focused.

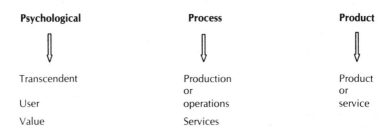

Figure 4.1   The quality views.

4.3 The five quality bases

Garvin (1988) suggests there are five bases of quality:

1. Transcendent.
2. Product based.
3. User based.
4. Manufacturing based.
5. Value based.

The aim here is to evaluate these views openly. Each is seen as distinct and separate and perhaps in varying terms mutually exclusive.

4.3.1 *Transcendent quality view*

Pirsig (1974) wrote, 'Quality is neither mind nor matter, but a third entity independent of the two. . . . It cannot be defined, you know what it is'. Using this approach, quality is totally personal and can escape definition, even by you. Garvin (1988) suggests that quality, using this view, 'is a simple, unanalysable property we learn to recognise only through experience'. It is something we cannot touch, but know instantly and can differ, over time, in relation to the same thing. Because quality has a personal feel here, Tuchman (1980) considers that quality is related to fine workmanship rather than mass production. Therefore, an individual will project a personal, subjective view of quality onto an object, for example, as long as the object provides the reinforcement that they have learned previously. When this reinforcement stops, the quality view changes and the projected quality view of the object is lost.

The importance of this view is not altogether understood, because until now there has been little research in this area. However, it is contended that a consumer who purchases a product or service does so because the quality view developed under this premise, is reinforced positively by that purchase and consequent use.

4.3.2 *Product-based quality view*

To change customer needs into defined terms of generating customer requirements (customer ⇒ design), and then customer specifications, marketing generally uses a product/user-based strategy. Thus, a product-based design strategy where quality is determined as a *precise* and *measurable variable* (Garvin, 1988) and differences in quality thus reflect differences in the quantity of some ingredient or attribute seen to be possessed by a product.

Products provide the basis for this quality view, as a function of the real features of the product, and quality is seen to rest solely with the product and not with the individual. However, changes in the individual's viewpoint change the acceptance of those features, and we are at once back to the previous approach and viewpoint.

4.3.3 User-based quality view

In a user-based strategy, definitions are based on the premise that quality is determined solely by the user. Individual consumers are assumed to have different wants or needs, and goods that best satisfy their preferences are the ones they regard as having the highest perceived quality. This therefore reflects a highly personalized and subjective view. Again, this only reflects the general *market segment*, not the individuals within that market.

4.3.4 Manufacturing-based quality view

To change design specifications to product output, manufacturing would generally use a manufacturing strategy. This is where engineering and manufacturing processes are specifically considered. This also suggests where, say, Crosby's *conformance to requirements* can be seen to be derived. The manufacturing strategy seeks to ensure that the deviations from the standard set — design specifications — are minimized, as deviations are seen to reduce the quality of the product so produced. This applies to services such as education. This does not mean that the product is inferior, but does mean that the quality that resided in the design specification has not been met. Thus the focus is internal, i.e. to the design specification. If the design specification is indifferent to customer needs, or if the design specification contains specifications that the manufacturer cannot hope to satisfy, then it is seen as a symptom and highlighted weakness of the design process. It is thus a product of design, not manufacturing. A trade-off in the development of the manufacturing strategy is that improvements in quality lead to lowering of overall product costs over time. So this strategy is characterized by increasing quality (lower deviations) by focusing on lower costs.

4.3.5 Value-based quality view

Here, Garvin (1988) suggests that the basis for this view is the psychological understanding of the meaning of *value*. It is an independent determination that reflects an individual's cost bias. Garvin's discussion includes the point that a $500 running shoe is not a quality product because it would find few buyers — but this is not necessarily so.

This value-based judgement actually reflects a manufacturing-inspired view, from the days when products were bought according to grade, rather than anything else. Consequently, consumers have been conditioned to accept that the 'quality' of a product is determined by price. This view is entrenched in Western society today. Even a study by The Consumer Network (1983) suggests that 'quality is apt to be discussed and perceived in relationship to price'. Therefore to many people quality is defined in terms of price, where a low price means low quality, etc. So why are manufacturers trying to develop a pricing strategy of low price and a 'high' quality product? This seems to demonstrate a paradox that has really not yet been resolved.

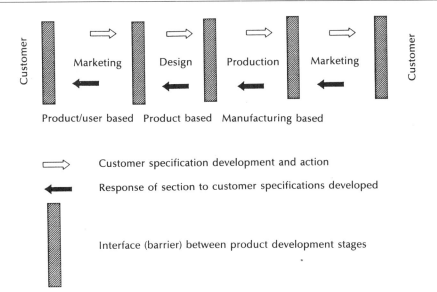

Figure 4.2 A customer–supply model.

4.4 Where does the customer fit into all this?

We shall use design requirements for products/services as an example of understanding where the customer fits into the three bases of quality.

Traceability of design requirements from customer through to supply, requires continuous information interactions, where later design outputs are checked against design specifications previously set. In practice, this becomes hierarchical in nature, where higher level specifications direct the more detailed design outputs. However, data transference through the hierarchy requires the effective crossing of the various interfaces that are deemed to exist between the customer and supplier. To illustrate this, consider Figure 4.2.

Between each section of the product development, there is an interface that needs to be crossed in order for the design specifications on the left to become the design product on the right.

4.5 The effect of the differing views of quality

There is one major problem. Traceability — the need to demonstrate the customer's voice throughout the design, production and service arrangements — requires the continuous and successive perceived quality of the customer needs, through to purchase by the customer. To change customer needs into defined terms of generating customer requirements (customer ⇒ design), and then customer specifications, marketing generally uses a product/user-based strategy.

This is a product-based design strategy, where quality is determined as a precise and measurable variable and differences in quality reflect differences in the quantity of some ingredient or attribute possessed by a given product.

In a user-based strategy, definitions are based on the premise that quality is determined solely by the user. Individual consumers are assumed to have different wants or needs, and goods that best satisfy their preferences are the ones they regard as having the highest quality. Therefore this reflects a view that is highly personalized and subjective. This has given rise to precise combinations of product attributes that generate the most satisfaction with customers within a specified market. Again, this only reflects the general market, not the individuals within that market. To change the customer specifications into design specifications, design engineers would generally use a product-based approach, as described above.

To continue, to change design specifications to product output, manufacturing would generally use a manufacturing strategy. This is where engineering and manufacturing processes are specifically considered. The manufacturing strategy seeks to ensure that the deviations from the standard set — design specifications — are minimized, and that deviations reduce the quality of the product.

What we have essentially discussed is that there are within an organization competing views on quality and these views need to be placed in a design context without compromise. This means that the generated views reflect the cultural determinism of each group and that compromise is very difficult when dealing with cultural issues as deep as quality views. Therefore, to ensure that all views are strategically determined, a method or methods need to be adopted or developed to ensure traceability of design requirements from customer/marketing research through to supply.

4.6 Factors affecting customer perceptions of quality

When dealing with the factors that are considered to affect how a customer perceives quality, it is possibly prudent to evaluate Garvin's (1988) eight dimensions of quality. These are:

1. Performance.
2. Features.
3. Reliability.
4. Conformance.
5. Durability.

6   Serviceability.
7   Aesthetics.
8   Perceived quality.

According to Garvin, these dimensions are independent and quite distinct. They may however also be interrelated, i.e. durability and conformance. In one product, a dimension may be critical to its success; in another, the same dimension may not be considered as crucial. The dimensions therefore provide a basis for evaluation of the characteristic elements of any product/service and therefore need to be interpreted widely as a consequence.

4.6.1 *Performance*

This includes the *primary* operating characteristics of the product or service. In relation to the bases of quality it would mean the application of product- and user-based notions. Examples include, for a car, the acceleration, miles per gallon, etc.; for a radio, the scope of transmissions it can receive, etc. The relationship between performance and quality is simple on one side to understand, but dangerous to accept in today's environment.

Customers have been conditioned over the years to accept that to get better performance required the development of better and superior products, which cost more. Consequently, this relationship (according to the conditioning) means that if the customer pays more for a product they will get enhanced performance in return. This kind of thinking now needs to be severely questioned, and the adoption of a customer orientation should be developed not only by the producer, but also by the customer.

4.6.2 *Features*

These are the *secondary* characteristics that supplement the product's basic functioning. These provide the additional set of attributes that contribute to the *whole* package that the customer buys. This element provides for flexibility in dealing with customers and therefore can be seen as a competitive weapon if used effectively. For example, a car is just a car, isn't it? For cars of perceived equal performance, it is these characteristics that help differentiate the product in the mind of the customer. However, Garvin warns that making a distinction between features and performance is sometimes difficult.

4.6.3 *Reliability*

Reliability is the function of a product/service to perform as expected over a specified period of time. It is generally measured using mean time to first failure (MTFF) and mean time between failures (MTBF), although other measures are used (Juran, 1974). The concepts can be applied equally to products or services, but their use is generally restricted to products, especially durable products. The critical nature of some products, e.g. aircraft engines, means that reliability characteristics are highly regarded. This is especially true when the costs of down time and maintenance are relatively high.

As quality management becomes more widespread, the application of techniques that ensure reliability in products and services will increase and become the norm rather than the exception. This will mean that customers will expect this characteristic and think nothing further. This is further emphasized as many household products do not contain user-replaceable parts and therefore the cost of repairing these items, in the event of a breakdown, increases correspondingly.

4.6.4 *Conformance*

Conformance is the degree to which a product's design and operating outcomes meet the developed standard. Conformance is the central theme in quality management. The definitions of quality by the various gurus all adhere to this concept. It is also the main theme in the application of quality management standards such as BS EN ISO 9000. This is also the main element of the Japanese quality management methods. It is through these developed techniques of measuring conformance that the Japanese have become so superior across the world's markets. It is considered that there are generally two ways of measuring conformance.

The first approach reflects the evaluation of what is being produced to a standard — the manufacturing techniques of process control and sampling. Many of these techniques use the approach of acceptable conformance as long as it is within specification limits (otherwise known as tolerance) — as laid down in the standard.

As Garvin points out 'there is little interest in whether the centering dimension has been matched exactly'. This means that exact tolerances are not called for and specification slack is acceptable. The work of Crosby (1979) on zero defects reflects an approach that seeks to eradicate this type of method. The second approach is embraced in the work of Taguchi. He developed the notion of the loss function — the losses that are imparted to society after the product is delivered (discussed in Chapter 3). Taguchi's method is to evaluate the cost of the variability around a target (the centre of the specification limits in the previous approach). The two differ in outcomes but use similar input data. Comparisons can be made, but are hampered by the more mathematically complex requirements of Taguchi's method. Thus, Taguchi's method lacks effective application on the shop floor, as a consequence.

Conformance and reliability are related in the manufacturer's quality view. It could also be said that reliability rests firstly on the design specifications being correct and secondly the conformance to those specifications.

4.6.5 *Durability*

Durability is seen as a measure of a product's life. Services have little or no durability. There seem to be some problems in determining an effective definition of durability. Do we mean the life of product before we must replace it? Or do we mean the life of a product before strategic parts of that product are replaced? Alternatively, do we use the technical capabilities or the economical aspects as the deciding factor?

Within a technical sphere, durability reflects the length of time the product can be effectively used, given the constraints of the user requirements (which the manufacturer should have considered anyway). Where a product cannot be serviced by the user, the choice the consumer has is generally whether to make the purchase again.

Where the product has consumer-serviceable parts, the choice is increased to include repair and it is here that the consumer may make mistakes. In this instance, durability of the product is measured in terms of warranty, repair, down time and replacement costs. The measurement also includes the transcendental dimension related to fashion, taste and status.

Durability and reliability are seen as related issues here. The use of guarantees to customers will give them greater confidence in the product offered than those of another company which does not.

If we look at customers' behaviour, products with increasing life spans, e.g. cars, may actually reflect the economic situation prevalent, rather than increases in technical product capability. As Garvin suggests 'durability is a potentially fertile area for further quality differentiation'.

4.6.6 *Serviceability*

Serviceability is the ability to offer a resumption in the normal working pattern of the user. This means the use of speed of service, availability, lowered cost and the effective development of a professional relationship between user and supplier. This element closely ties in with other dimensions such as, reliability and performance. This is one dimension that is clearly visible to the consumer and much attention has been given to this area in recent years. This applies to manufacturing and service sectors alike. From the manufacturer's viewpoint, the proactive development of quality policy and procedures to assist in this area are becoming a necessity, rather than something that is attended to as the need arises.

4.6.7 *Aesthetics*

This depicts the consumer's response or reactions to characteristics such as, touch, taste, smell, looks and sounds. It is individual in nature and reflects personal judgement. What looks good to one individual does not look good to another, even though the performance, reliability and durability characteristics are the same. What is missing is that the *total* conformance characteristics are different. It is a powerful dimension. Although individual in nature, it can be seen to reflect group norms and biases, i.e. fashion.

4.6.8 *Perceived quality*

This reflects the notion that consumers buy products/services with incomplete information about the total characteristics that make up that product or service.

Consequently, the consumers' information gained through dealing directly with the supplier, knowledge of similar products and knowing exactly what it is they actually want from the product or service, give them some measure on which to evaluate. It is this evaluation that leads to the notion of perceived quality, and is totally personal.

Perceived quality is perhaps the only element that overrides all other elements in the buying process — at least initially. This means that reputation — stated or implied — has a powerful influence on the psychological development of perceived quality.

4.7 A comparison of the five quality bases and the eight dimensions

A comparison can be found in Table 4.1. The quality bases relate quite well to the dimensions of quality. The first five dimensions are closely connected to the manufacturing process, and the final three relate to the individual. What is missing here is the development of a coherent connection between dimensions and the quality bases. Although Garvin has indicated that the dimensions are distinct, they are nevertheless related in some way. How they are related signifies the basis for the development and application of the *quality management mix*.

Table 4.1 Detailed activities for quality improvement.

Dimensions	Quality bases				
	Transcendent	Product	User	Manufacturing	Value
Performance		*	*	*	*
Features		*	*	*	*
Reliability		*	*	*	
Conformance			*	*	
Durability		*	*	*	*
Serviceability	*	*			*
Aesthetics	*				*
Perceived quality	*				*

4.8 Chapter review

In this chapter we have explored the basis for the meaning of quality. More specifically, Garvin's viewpoints have been analyzed and evaluated. It is important to understand that none of these bases is more influential than any other. However, it has been demonstrated that the manufacturing quality (conformance, predominantly) has been taken as the leading quality base. This is changing as the needs and wants of the consumer have taken on greater significance, and influence the manufacturing operations much more openly today.

The quality views in themselves are powerful windows in which to evaluate their importance. It should not become admissible to accept only one view, as had occurred in the past. A product will sell to a pluralistic audience. Consequently, until the science of evaluating customers' needs and wants is developed to the stage where there is no doubt as to their requirements, then the other quality bases will offer significant influence on the product/service development.

4.9 Chapter questions

1. Compare and contrast the five bases of quality. Evaluate the effect of the differing views of quality.
2. Discuss the relative position of the customer — product/service, in relation to these quality bases.
3. Discuss the eight dimensions of quality.
4. Compare and contrast the five bases of quality with the eight dimensions of quality.

Part 3

The Five Functions of Quality Management

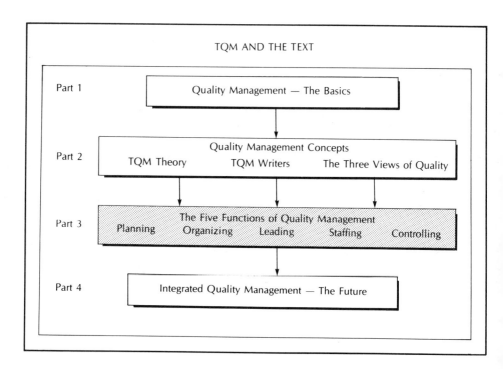

CHAPTER 5

Quality planning

CHAPTER OBJECTIVES

Explain the reasons and need for planning for quality
Describe the various types of quality plans
Evaluate the meaning of quality planning
Describe the quality planning process
Discuss the various types of benchmarking
Describe the benchmarking process
Evaluate the benefits and implications of benchmarking

CHAPTER OUTLINE

- **5.1** Introduction
- **5.2** Why plan?
- **5.3** Why the need for planning for quality?
- **5.4** Who is responsible for planning for quality?
- **5.5** Types of quality plans
- **5.6** What is quality planning?
- **5.7** The quality planning process
    - 5.7.1 *Environmental analysis*
    - 5.7.2 *Quality mission*
    - 5.7.3 *Setting a quality policy*
    - 5.7.4 *Generate strategic quality goals*
    - 5.7.5 *Establish quality action plans*
    - 5.7.6 *Quality strategy implementation*
    - 5.7.7 *Monitor and evaluate quality performance*
- **5.8** Benchmarking
    - 5.8.1 *Definitions of benchmarking*
    - 5.8.2 *Goal setting and benchmarking*
    - 5.8.3 *Types of benchmarking*
    - 5.8.4 *Characteristics or indicators used in developing benchmarking practices*

              5.8.5 *Influences on the benchmarking process*
              5.8.6 *The benchmarking process*
              5.8.7 *Benefits of benchmarking*
              5.8.8 *Benchmarking limitations*
5.9   Chapter review
5.10 Chapter questions

5.1 Introduction

Planning is the first function of management. It is future-oriented and provides the direction for the whole organization, its importance cannot be overstressed. If you do not get your planning right, you will waste resources, energy and reputation chasing fruitless exercises destined for failure. Oakland (1989) indicates that systematic planning is a 'basic requirement of TQM'. The intention here is to provide a framework for the development and evaluation of quality-related planning issues. The first step is to determine the managerial context of quality planning.

5.2 Why plan?

Planning by definition means preparation for change. Essentially, for most businesses, this means change, engineered every twelve months or so. This change coincides with the need to plan ahead for the new, yearly, budgeting round. Planning in this context means:

1. Evaluating the past and present in order to ensure the best possible future.
2. Determining the most objective and prudent course of action, given those circumstances found in (1) above.
3. Carrying out that action and monitoring results.

As a consequence of the above, planning should lead to improved performance. Its advantages include that it:

1. Helps management determine their strategies for adaption and correlation to changing environments, both internally and externally.
2. Develops courses of action that provide some element of consistency that managers, staff and customers can appreciate.
3. Provides information so that meaningful decisions can be made.
4. Helps to ensure co-ordination between inter- and intra-elements of the organization.

Its disadvantages include:

QUALITY PLANNING 95

1. Planning for the sake of planning, not testing the outcomes effectively.
2. Benefits accruing from the planning outcomes do not exceed the expense (in human as well as financial or resource terms) of developing the plan.
3. Planning creates delays to intended implementation schemes.
4. Planning reduces creativity, spontaneity and innovation.

The need for effective planning should be foremost in every manager's mind and the need to ensure that all staff understand and can apply the principles of planning are deemed essential.

5.3 Why the need for planning for quality?

Planning brings to decision making the capability of being proactive and thus anticipate future events and the necessary actions to meet those events positively. Ivancevich *et al.* (1994) indicate three reasons why planning is necessary:

1. Decreasing timespans from decision to results becoming available from those decisions — managing this shortening timeframe effectively is becoming the key to ensuring that organizations get to market first with a product/service.
2. Increasing organizational complexity — heightened requirements for international co-ordination developing out of the need to use ever-increasing advances in technology.
3. Increasing international competition — customers are no longer forced to accept what is on offer from the *home* manufacturers and consequently international competitors force lower prices and generally enhanced products/services. These pressure local manufacturers to follow suit.

5.4 Who is responsible for planning for quality?

In classical strategic planning, top management has this responsibility. It is centralized in form and operation. Divisions or departments have the responsibility to implement the results of developing the mission, objectives and strategies of the organization. This does not seem to be an appropriate mechanism for managing an organization which is fundamentally quality-oriented. Significantly, quality-oriented organizations develop and apply decentralized strategic planning practices. Again, this falls in line with the quality philosophy. However, even in these organizations, top management has generally reserved the right to develop the mission for the organization, while divisions or departments develop out of that mission their quality goals and strategies and implement them. Quality-oriented organizations make sure that their planning processes are effective by securing and emphasizing staff involvement, using resources — physical, human and financial — according to the quality plan and effectively implementing them.

5.5 Types of quality plans

Quality plans differ in terms of:

1   *Scope* — the range of quality activities covered by the plan.
2   *Detail* — the relative components of the quality plan.
3   *Timeframe* — length of time covered by the quality plan.
4   *Their application* (in terms of level).

It is accepted in general management theory that there are three types of plans. The same principles apply to quality planning. There are strategic plans, tactical plans and operational plans.

Strategic planning is an attempt to develop a long-term, comprehensive and prescriptive plan, suitable for the organization to understand where it is now, where it is going, develop effective goals to achieve and the strategy required to satisfy those developed goals.

Tactical planning is more contextual than strategic planning, but is not as focused as operational plans. It provides intermediate goals and direction for specific areas of the organization, e.g. marketing, design and production. These plans deal with effectiveness, making sure that the specific areas of organization are working on the right activities.

Operational planning is a short-term orientation, narrow, and a much more focused plan, that is seen as more objective than strategic plans in terms of measuring outcomes. The prescriptive nature of strategic plans is translated into realistic measures for performance. These plans deal with efficiency, making sure that the specific areas of organization are working on activities in the right way.

Table 5.1 indicates the nature of the above quality plans.

5.6 What is quality planning?

The systematic use of quality planning is vital to an organization's competitiveness. According to Juran and Gryna (1993), strategic quality management is the 'process of establishing long-range quality goals and defining the approach to meeting those goals'. The processes of general management theory can be applied in the same way for quality planning. However, quality planning provides a much more focused orientation to business strategy. Quality planning is essential for an effective and manageable quality improvement process. This highlights the difference between general planning theory and quality planning, and that is an

Table 5.1 Types of plans

Planning differences	Types of plans		
	Strategic	Tactical	Operational
Detail	General	Intermediate	Specific
Timeframe	Long term (3–10 years)	Intermediate (1–3 years)	Short term (0–1 years)
Application level	Organization-wide	Both	Departmental

improvement. Quality planning moves beyond strategic planning by adding a requirement for improvement. This is different to change, as all plans impart change by default. Consequently, quality planning has to be structured more effectively in the corporate managerial environment.

Gone are the days when a good strategic plan would ensure the survival of the organization. The age of quality is about delighting the customer and this cannot be done within a process that philosophically or practically does not improve. The quality plan should be seen as an extension of the strategic plan, not a replacement, at least initially. What is important is that management and staff work together to ensure corporate improvement, using the quality plan as the vehicle.

An interesting point to note, is that 'virtually all of the Deming Prize winners can point to clear, detailed, well-communicated total quality improvement plans' (Labovitz and Chang, 1990). (Note: The Deming prize was introduced in 1951 in Japan. Its purpose is to award prizes to those companies which are recognized as having successfully applied CWQC based on statistical quality control and which are likely to keep up with it in the future.)

5.7 The quality planning process

The general quality planning process consists of developing an organization's:

1. Environmental analysis.
2. Quality mission.
3. Setting a quality policy.
4. Generate strategic quality goals.
5. Establish quality action plans.
6. Quality strategy implementation.
7. Monitor and evaluate quality performance.

It becomes obvious that as you move through this process from the quality mission to developing operational quality goals, it also becomes more complex and much more difficult. The reason for this is the translation must take into account the development of essentially prescriptive elements (quality mission) to the essentially practical orientation of the operational goals. The outcome of this is a quality plan that encompasses all that the organization wants to achieve, how it is going to achieve them and by when. It is a document that provides visible support for the intentions of the management of the organization.

5.7.1 *Environmental analysis*

An environmental analysis is required to ensure that what customers want, they get. To do this, a systematic search — both internal and external — is required. This involves carrying out a SWOT analysis (Strengths and Weakness — both internal; Opportunities and Threats — both external).

In relation to the types of plans that need to be developed:

1. General strategic issues — macro issues — such as sector/industry trends would be evaluated. These include long-term changes to industry technologies, workforce requirements and capabilities, legal, environmental and economic issues. The scope is in line with the considered timeframe of the strategic plan.
2. Operational plans would require the analysis of much shorter timeframes. These would require investigation into operating performances and the effects of shorter-term issues affecting those performances. These would include union and workforce issues (job requirements, training and remuneration), individual and group performances, and operating performance issues such as cash flow, wastage, inefficiencies and operating costs.

5.7.2 *Quality mission*

A crucial element is the development of the quality mission. This gives direction to the organization and an indication to all stakeholders of what is important to management. The quality mission statement should be in written form and short, clear and concise. If it is implicit in the organization because that is the way it is done here, then there is a danger of individuals thinking they are carrying out the mission of the organization, when in fact, they are not. The written statement is about ensuring the visibility of the hand of management.

In generic terms, the mission is the fundamental reason or purpose of existence for the organization that distinguishes itself from others. In relation to quality, this is translated into the fundamental reason for operating. The quality mission therefore supports the overall mission for the organization. It also provides the framework in which all organizational activities are correlated. It therefore has a strategic role of guiding the organization through the use of quality policy.

The mission statement can be viewed as a benchmark (Bartol and Martin,

1991) — what the organization intends itself to be — for strategic targeting and evaluation by both managers and workers alike.

It is an important icon for all staff, but more importantly for staff other than management, because they are generally not privy to the information and environmental context in which the statement was developed, formulated and approved.

An example of a mission statement is 'the provision of a structured learning environment, directed at undergraduate international business education, whilst focusing on developing cultural interactions that benefit the European and wider community' (an educational department in a university).

5.7.3 *Setting a quality policy*

One of management's responsibilities is to set policy. The policy should relate to their commitment and positive belief of the quality philosophies, principles and practices. The quality policy is the first substantial visible evidence that management is serious about what they want to achieve with quality. However, quality policy:

1   Gives guidelines of *what* is to be done, rather than the actual *how* of doing things.
2   Is generic in operation and applies to whole classes of schemes of work.
3   Is prescriptive, but should have the power to assist, rather than hinder, quality-related performance.
4   Applies organization-wide.

The quality policy serves as the integrating factor that quantifies the mission as guiding principles. One major effect is that the policy will be scrutinized by both internal and external agencies. This means that the organization will have to deliver what it promises or risk affecting its survival. Not all effective policies are written down, though. To illustrate this point, Nuland (1990) indicated that the General Manager of UCB Chemical Sector insisted that all staff members receive quality management training. This is the necessary and expensive commitment of top management, but nevertheless can illustrate the serious intention of top management, formulated through the quality policy.

When developing the quality policy there is a need to consider the following, *inter alia*:

1   Who, what and where are the customers?
2   What products/services do they require, and when?
3   What are the competitors' intentions and what does their quality policy indicate?
4   What is the focus of the quality mission?
5   Who should be involved in developing the quality policy and who is going to lead its formation?
6   Should the supplier(s) be involved?

The overall approach is to develop a policy that can be accepted organization-wide. This development will not preclude the formulation of localized quality policies at, say, department level. On the contrary, this should be encouraged, as it leads to further involvement of staff at the point where it matters — operations.

Policy issues and implementation may create problems for staff though. The nurturing of a positive work environment is crucial, especially when a policy seems to work against a section of the workforce, e.g. policies directed towards reducing waste may mean reducing costs in rework. This may result in fewer people working in this area and redeployment may be required. This brings up the point about *not* developing policies in isolation in order to reduce these perceived or real problems, but developing a cohesive policy that is positive in the long run, by making sure that the approach is integrated.

5.7.4 *Generate strategic quality goals*

A goal is a target to be achieved. A goal needs to be:

1. Determinable.
2. Actionable.
3. Measurable.
4. Specific, e.g. time.

Juran (1988) suggests that organization-wide quality management is a 'systematic approach for setting and meeting quality goals throughout the company'. A question to pose is, How are quality goals established in the first place? There are a number of methods that depend on the operating circumstances. These include:

1. *Past-performance data* — This is where historical data have been collated to provide the basis for the quality goal's evaluation. This is used greatly in the manufacturing or engineering areas of the operation. It is now becoming more popular with the service sector. The advantage of using this type of data is that people are more comfortable using their own generated data.

    This may preclude an objective analysis, simply because they may be collecting the wrong type of data for the future needs of the organization, and this needs to be scrutinized closely. Examples include the use of production requirements — material used, inventories, workforce hours, scrap and rework costs, etc. In the service sector, these could include service performance parameters such as numbers of clients served, types of service required, etc.

2. *External environment* — Setting quality goals based on the external environment (customers) parallels the quality philosophy. The changing behaviour of the external environment means that information about its

tastes, requirements and buying behaviour is becoming more important. Consequently, developing quality goals that do not match this will lead to an ineffective quality strategy. This area is becoming more important as, for example, benchmarking is based on evaluating for any given sector, or competitor organizational quality accomplishments.

3   *Standards setting* — This is where a specification is developed for a product or service after specific and concentrated evaluation. The approach is particularly well used in engineering and manufacturing areas, where new and innovative products are developed. The specification provides the standard to achieve and the outcome, the basis for the development of quality goals.

4   *Competitor based* — This means evaluating the products/services produced by competitors and re-engineering your own processes by developing appropriate quality goals as a result. Evaluations can take the form of physical test and inspection — especially products and actual field performances — especially services.

The above are essentially methods that seek to use previously developed data, and use these data to make meaningful decisions about the future. The use of more than one method will ensure that there are more objective data on which to develop those crucial goals.

Inputs to quality goals can be systematic (as in the above) or spasmodic, as in the case of suggestions from staff. They can also come from the application of quality techniques such as Pareto analysis, cause and effect diagrams and SPC, and from legislative spheres.

Whatever the source, and the extent of the data's credibility, any quality goal developed needs to be *sold* to the staff. This can be a very difficult process to achieve. Staff are more likely to accept quality goals that they have had an opportunity to help in the formulation. In this regard, strategic goals would need to be broken down into divisional and/or departmental goals in order to account for this.

Juran and Gryna (1993) indicate that quality goals can be used for breakthrough or control (see Chapter 3). These are for a breakthrough:

1   They wish to attain or hold quality leadership
2   They have identified opportunities to improve income through superior fitness for use
3   They are losing market share through lack of competitiveness
4   They have too many field troubles — failures, complaints, returns and wish to reduce these as well as cut external costs
5   They have a poor image with customers, suppliers, etc.

These goals are related to an ongoing need for quality improvement. Once they have been developed, they have to be implemented in the areas that management perhaps already senses that they have problems. Consequently, the development of these goals for breakthrough provides an opportunity to develop the business process and possibly more importantly, the staff. Management commitment is tested further, as quality goal development goes beyond the rhetoric, and tests and challenges the practicalities of organizational operations.

5.7.5 *Establish quality action plans*

Actions are similar to strategies, they formulate the planned courses or strategies needed to be adopted to accomplish quality tasks. In this book they are differentiated by timeframe and scope, i.e. strategies are normally developed for longer timeframes and wider scope than actions. However, actions may be seen as more important, because they form the base on which the strategies rest in terms of implementation and evaluating outcomes. Both these elements must be continually directed to providing the solution to the objectives sought. In this respect, there may be different actions that can provide the solution to the set objective and consequently management will need to make a choice as to which alternative gets implemented.

Action plans are therefore necessary to ensure continuity of the quality plan. They directly relate to implementation issues and outcomes, and will account for the detailed treatment given to them. Without action plans, quality plans will lose their ability to provide a cohesive planning force for the organization. It is also the place where subordinates can wield great influence on the effectiveness of the quality plan.

5.7.6 *Quality strategy implementation*

One of the most important stages in the quality planning process is the implementation stage. Issues of implementation include:

1. *Education and training* — Directing the quality plan to individuals who have been given little or no quality education/training must be avoided. Individuals cannot be expected to take on board new job and task requirements and to perform effectively without adequate education and training.
2. *Participation* — Individuals who have participated in the development of all quality actions and quality-related plan development — appropriate for their level — will be likely to implement the quality plan more effectively by being more committed.
3. *Cultural* — These issues surround the need to enhance the change of behaviour for new plans. Since quality plans are future-oriented, then changes will be likely, and therefore the expectancy for individuals to make appropriate changes also.
4. *Technological* — As quality plans consider both the *what* and *how* of implementation, issues surrounding differing and changing uses of technology will become apparent.

5   *Process* — Quality planning will effect changes in the processes that an organization uses to implement its quality plans.
6   *Authority/power* — The quality-oriented organization seeks to ensure that the authority/power base is given to those who need it — the implementers.
7   *Reward structures* — These may need to be modified, not just because of changes in expected performance, but also for changes in responsibilities. Consequently, reward structures will need to reflect the changes to organizational quality practices. Autonomous groupings implement plans with greater responsibility. Rewards include bonuses, salaries, promotions and the development of skills and knowledge.
8   *Organizational structures* — As quality practices have historically resulted in flatter organizations, the quality-oriented organization will need to address issues of squeezed middle management and enhanced managerial roles of lower staff.

Implementation — the doing — is different from the final stage in the quality planning process — monitoring and evaluation of quality performances. If sufficient planning has not taken place before the implementation phase, then all the work carried out previously could be wasted. Nor does implementation mean just doing. Since the implementation phase is acting at the leading edge for quality development in the organization, it makes good sense to monitor it continuously. This means monitoring with a low level of scope and timeframe, providing flexibility and opportunities for innovation, at the point where it is most fruitful.

5.7.7 *Monitor and evaluate quality performance*

Quality plans require continuous monitoring in order to ascertain their effectiveness. This would mean the development of monitoring systems at all levels in the quality plan. It means the reverse of quality plan implementation, i.e. quality plans are implemented downwards, whereas quality plan monitoring and evaluation are accomplished upwards. The crucial element is the day-to-day generation of data. This was not the focus in traditional planning; in quality planning, it is. The performance data thus generated, using quality tools, provide statistical as well as real measures of quality performance.

Monitoring should be used to effect strategic and operational quality plan evaluations. These check the viability and effectiveness of the designed quality plan against the implementation outcomes. Developing organization-wide quality information systems will ensure that co-ordination, competence and continuous implementation issues of the quality plan are addressed.

The application of a quality audit will need to occur in this phase. Quality audit in this instance means an independent and formal review of quality-related performances of the quality plan. The assessors will invariably be individuals who have no line responsibility for the outcomes of the quality plan, but who have the authority to demand information that will allow them to make an effective evaluation of quality performances. The outcome of the quality audit will in itself be subject to continuous improvement. This would evaluate both the content of the quality plan and the process of evaluation.

5.8 Benchmarking

5.8.1 *Definitions of benchmarking*

The expressions and meanings attached to benchmarking first originated from surveying. However, the modern term bears little resemblance to this activity.

Bemowski (1991) suggests that benchmarking can be defined as 'measuring your performance against that of best-in-class companies, determining how the best-in-class achieve those performance levels, and using the information as the basis for your own company's targets, strategies and implementation'. The donor process is the superior process that is being evaluated in order to provide model performance data for continuous improvement of the recipient process. Benchmarking does not seem to be limited to the singular process under examination (the recipient process), because the newly developed data (from the donor process) must be used to effect an enhancement in the recipient process, the resourcing of that process, and the organizational strategies that are outcomes of the developed objectives for that process. Durgesh and Evans (1993) suggest that benchmarking is a 'continuous, structured process that leads to superior performance and a competitive advantage' and that the key to benchmarking is not to focus solely on numbers, as this is deemed ineffective, but there must be a focus and a commitment to process improvement. It is structured because it is a planned activity designed to improve continually. Shetty (1993) suggests benchmarking outcomes must not just match the competition, but must outperform it. However, benchmarking just because it is the in thing to do, may not provide the enhanced performance required.

One of the critical benchmarking requirements is the identification of a superior process that will provide the necessary outcomes and developed data for the application of continuous improvement. Continuous improvement can only be adequately addressed though, if the evaluated process or processes are continually compared (Main, 1992).

This cannot be guaranteed. Consequently, the choice of process requires prior knowledge or inside information about its relative performance(s). Much of the power of benchmarking is lost if a donor process does not yield adequate performance improvements after application.

The promise of benchmarking does not seem to be brought to fruition unless essential elements are developed and committed to. Pryor and Katz (1993) suggest that these elements include:

1. A commitment by management with visible support.

2   Detailed planning and plans.
3   Constant communication throughout the organization.
4   Adequately developed teamwork.

Here, benchmarking means:

1   Determining appropriate characteristics of the recipient process to be used to compare one process with another (donor).
2   Developing data about the performance of a best-practice process contained within or outside of the organization requiring benchmarking data.
3   Comparing and evaluating the process or processes against the data relating to the characteristics measured.
4   Developing continuous improvement measures from the new data.
5   Implementing the planned changes to the process.
6   Monitoring the effectiveness of such changes.

Benchmarking therefore requires a planned action of evaluation and implementation. It is an attempt to modify a process, in the light of new knowledge gained about a more effective process. TQM is judged as internally oriented, with continuous improvement at its core, whereas benchmarking can be seen as externally oriented, with continuous improvement only if the process benchmarked results in an improved process evaluation.

5.8.2 *Goal setting and benchmarking*

Benchmarking requires the development of measurable characteristics that are reasonably available in both the donor process and the evaluated process. These characteristics translate into organizational goals only after successful benchmarking practices have been identified. Also, these successful characteristics can only be used to make changes effectively if top management support their implementation. Benchmarking therefore requires that top management commit themselves to the outcomes of the efforts of benchmarking teams, when they clearly determine measures that can reflect improvement in processes when implemented.

Shetty (1993) suggest that 'successful firms use benchmarking to be creative, not reactive'. This underpins the need for planning benchmarking practices. However, the results may need to be implemented on an ad hoc basis, as each evaluation may not provide the necessary data to make the effort of implementation worthwhile. Vaziri (1993) recommends that a manager contemplating benchmarking should ask seven questions about 'who and what to benchmark, what to measure, how to collect data, and how to implement what is learned by the benchmarking initiative'. Answers to these questions therefore provide a ready-made basis for evaluating whether the benchmarking activity is worthy of the effort involved and the proposed outcome changes to process performance. Vaziri also suggests that the developed data need to be analyzed and any positive influences should be used to change the goals relating to the recipient process. In this way, changes caused by knowledge of performance gaps can be directly focused on the outcomes of the recipient process. This, however, has further implications for the scope of such changes. Care must be taken to ensure that process objectives are not changed in isolation and that any resulting actions are part of an integrated strategy.

The organizations which have most to gain from the careful application benchmarking practices are small ones. Generally, they do not have the wealth of experience of larger ones. Consequently, benchmarking can be seen as a means to reduce the learning curve considerably. No longer do companies need to feel isolated, because they can develop partnerships in developing effective performances that create a competitive advantage.

5.8.3 Types of benchmarking

Benchmarking can be subdivided into three areas:

1. *Internal* — An evaluation of practices within one organization. It is a method where knowledge about the processes is uncovered, usually by members of another department or group.
2. *Competitive* — Quite limited in realistic application, as it requires competitors to acknowledge and co-operate in order to assist one or both organizations to improve.
3. *Cross-industrial* — Evaluations between operations in different industries. Leads to the adoption of integrated cross-industry generic competitive practices.

5.8.4 Characteristics or indicators used in developing benchmarking practices

The *what* of benchmarking determines that it is necessary to understand all there is to know about the home process and its performance (using flowcharts), that will be used as the recipient focus for the comparison. Consequently, a plan of action is needed that uses the home process as an evaluative basis.

Defining the processes that are critical to competing and improving critical performance metrics (Monczka and Morgan, 1993) that relate to the recipient process is vital. Developed performance indicators link the processes and provide an evaluation of gaps and the causes of those gaps and inadequacies. Over time, this results in a spread of performance outcomes and trends and their consequent causes. It is this very nature and application that provides benchmarking as a necessary and effective quality management tool.

Training divisions may compile benchmarking metrics such as measures of training activity, measures of training results and efficiency (Ford, 1993). Ford also indicated that training benchmarks were easily compiled and evaluated, as was demonstrated at three Malcolm Baldrige Quality Awarded organizations.

Benchmarking means the adaption, rather than the copying of best practices. It means using the knowledge of your process in order to determine what is usable from the donor process. In this way, the mentality or culture surrounding benchmarking must be to improve, to exceed the performance dimensions of the donor process. This is why TQM and benchmarking go hand-in-hand.

5.8.5 *Influences on the benchmarking process*

The largest influence on the outcomes generated through the benchmarking process is the team that is sent to elicit the donor process data. Spendolini (1993) indicates that the team must contain functional expertise relating to the process being evaluated — communication skills and motivation. Camp (1989) explains that a team should contain three people — one to ask questions, one to take notes and another to think up the questions. This begs the comment that the action and need to think up problems on-the-spot signifies a reactive, rather than a proactive measure, as benchmarking must be. This practice therefore reduces the potency of the activity and should not be recommended.

The expertise of the team members is clearly a necessary aspect to consider, but what is perhaps forgotten is the impact of micropolitics both within the team and also in relation to where the team members *owe their loyalty*. The basis for this is that processes taken in isolation often affect the effectiveness of other processes in some way. Consequently, political overtones of the individual controlling those processes can affect the recipient process development and its outcomes. In many instances, the outcomes of such investigations are determined more on the balance of power, than on their technical nature.

Bemowski (1991) suggests setting teams containing six to eight people, of which, only three to four visit the donor process. This indicates that the other half of the team assists in the planning and development of the benchmarking activity, and helps to communicate and implement the outcomes of the donor process evaluation. As more people are involved in the whole benchmarking activity, there is likely to be greater commitment to accepting the developed changes. Therefore the benchmarking team should include key people from each department affected by the outcome of the benchmarking activity (Crow and Van Epps, 1993).

5.8.6 *The benchmarking process*

Various benchmarking processes have been developed, for example, Shetty (1993) and Camp (1989). The benchmarking process can generally be as depicted in Figure 5.1.

Again, continuous improvement is a requirement of the application of the process. Young (1993) suggests that the managerial process consists of four stages — planning, analysis, integration and action, indicating that some form of planned

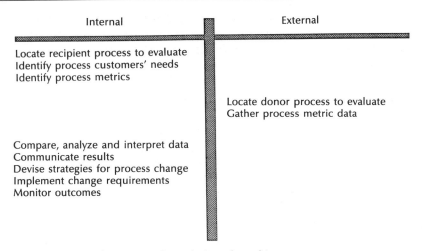

Figure 5.1 Generic benchmarking process.

commitment is required. Brelin (1993) further indicates that the process includes planning, data collection, analysis and adaption to the new information. What is missing here is Camp's (1989) fifth phase, maturity, which determines that an organization in this phase is not successful until all the best-known industry practices are incorporated into the organization's processes. However, benchmarking practices would now force the organization to go beyond the home industry and seek out processes for mirroring and improvement.

5.8.7 *Benefits of benchmarking*

The major benefit to be had from the successful implementation of benchmarking practices is enhanced process performance, resulting in fewer errors, less waste, knowledge of the process and its present and potential capabilities. Haserot (1993) suggests that benchmarking results in 'a rational method to set performance goals and to gain market leadership and a broader, more accurate organisational management perspective'. It provides these lofty benefits only if the organization is truly committed to the benchmarking activity and has judiciously applied them throughout.

The application to one major — but critical — process cannot offset the inefficiencies found in the rest of the organization. This could result in wasted effort, not because of the new data and the commitment to implement the developed changes, but because the rest of the organization resists the implications of the expected change. Consequently, for the benefits to become real, an organization-wide process must be implemented.

Many writers discuss the benefits of implementing benchmarking practices. For example, Weatherly (1993), Bogan and English (1993) and Vaziri (1993) suggest that the benefits of benchmarking include:

1. The provision of numeric goals and indicators of relative performance.
2. The development of insights into innovative approaches by other organizations (departments) that affect process performance positively.
3. Generating a vision that is outward looking, while focusing internally in critical managerial issues and processes.
4. Top management know how their organizations fare against comparable organizations.
5. Assisting in the development of effective strategic management plans, organizational re-engineering, redesign and/or restructuring initiatives.
6. Supporting a learning culture that values continuous improvement.
7. Focusing resources on developed performance targets that relate to demonstrated superior external performances.
8. Sharing information between process partners.

5.8.8 *Benchmarking limitations*

Although the basis for benchmarking is continuous improvement of a recipient process, the time, effort and value of the benchmarking process must be evaluated. As organizations become more effective, the increased performance achievement reduces, and thus further improvement is difficult to obtain. Consequently, intra-industry evaluations give way to interindustry evaluations. It is here, that benchmarking draws together many of the effective but generic techniques that can be used.

Hequet (1993) indicates that benchmarking to world-class performers is 'demonstrably helpful only to top-performing companies'. This would suggest that the choice of donor process must be relevant to the recipient process, not only in terms of required performance outcomes, but also the commitment or capabilities of the recipient organization. Hequet's research further suggests that presently low-performing processes may actually be *negatively* influenced, rather than positively — as a consequence of poor-quality infrastructures. This indicates that a recipient process cannot be treated in isolation, and that any planned improvements in that process, must be supported by the rest of the organization.

Other possible limitations could include:

1. *Fashion.* It is fashionable to implement benchmarking. Similar to the negative approach to quality management, implementing benchmarking because it is fashionable raises expectations of staff, without management becoming committed to its outcomes. Consequently, benchmarking loses much of its potency.
2. *Legal/ethical issues.* Notions of ethics and legal questions may need to be addressed surrounding the exchange of working information between organizations, especially competitors.
3. *Dependence.* Benchmarking relies on partnership dependence. This is the main resister to benchmarking competitor processes effectively, as much of the information is calculated to be confidential.

5.9 Chapter review

Planning for quality is the first function of quality management. Planning is future-oriented and provides the direction for the whole organization. Planning means evaluating the past and present in order to develop the most appropriate possible future, determining the most prudent course of action and monitoring its application and result. Planning decreases timespans for managers reviewing the results of decisions. Quality planning is the systematic establishment of long-, medium- and short-term quality goals and objectives and defining how best to achieve them.

The advantage of planning is that it provides information for managers to make effective future-oriented decisions' and helps to co-ordinate between departments. Some disadvantages include the reduction in creativity, spontaneity and innovation.

Quality-oriented organizations generally develop and apply decentralized planning mechanisms. This means that top management reserves the right to develop the mission and the departments and groups develop the quality objectives and implement them.

Quality plans differ in terms of scope, detail, timeframe and application. Three types of plans exist.

1. Strategic planning — Long-term comprehensive and prescriptive plans.
2. Tactical planning — Development of intermediate goals and direction for specific areas of the organization.
3. Operational planning — Short-term, narrow-focused plans.

The quality planning process consists of:

1. Establishing an organization's environmental analysis — Internal and external analysis using techniques such as a SWOT analysis, etc.
2. Developing a quality mission — Gives direction to an organization and indicates what is important to its management.
3. A quality policy — One of management's responsibilities is to set quality policy. It is an integrating factor that quantifies the mission as guiding principles.
4. The development of goals and action plans — Goals are developed mostly in a systematic way, but occasionally they can be spasmodic when receiving suggestions from staff. Goals are used to create direction and a means for measuring accomplishment. Action plans formulate the course of action needed to be adopted in order to accomplish tasks.

5   Implementing the plan while monitoring and evaluating quality-related performances — This includes dealing with implementation issues of education and training, participation, technological, authority and power and organizational structures. Continuous monitoring and evaluation are required in order to ascertain the effectiveness of the quality plans.
    Quality plans are implemented downwards, whereas monitoring is implemented upwards. The crucial element is the immediate attention to the generated data, which should be used to check the viability and effectiveness of the designed quality plan.

Benchmarking can be defined as measuring your performance against that of best-in-class companies, determining how the best-in-class achieve those performance levels and using the information as the basis for your own company's targets, strategies and implementation.

The benchmarking process means:

1   Determining appropriate characteristics of the recipient process to be used to compare one process with another process (donor).
2   Developing data about the performance of a best-practice process contained within or outside of the organization requiring benchmarking data.
3   Comparing and evaluating the process or processes against the data relating to the characteristics measured.
4   Developing continuous improvement measures from the new data.
5   Implementing the planned changes to the process.
6   Monitoring the effectiveness of such changes.

Benchmarking, therefore, requires a planned action of evaluation and implementation. It is an attempt to modify a process, in light of new knowledge gained about a more effective process. Benchmarking can be subdivided into three areas: internal — an evaluation of practices within one organization, competitive — quite limited in realistic application, and cross-industrial.

The *what* of benchmarking determines that it is necessary to understand all there is to know about the home process and its performance (using flowcharts) that will be used as the recipient focus for the comparison. Consequently, a plan of action is needed that uses the home process as an evaluative basis.

The largest influence on the outcomes generated through the benchmarking process is the team that is sent to elicit the donor process data. The expertise of the team members is clearly a necessary aspect, but what is perhaps forgotten is the impact of micropolitics, both within the team and also in relation to where the team members owe their loyalty.

Benchmarking limitations include the effects of fashion, legal/ethical issues and dependence.

5.10 Chapter questions

1 Evaluate the meaning of quality planning.
2 Describe the various types of quality plans.
3 Explain why planning is needed.
4 Discuss the reasons for developing a quality plan for an organization.
5 Describe the quality planning process.
6 Explain the benefits and limitations of benchmarking.
7 Discuss the process of benchmarking.

CHAPTER 6

Quality of design

CHAPTER OBJECTIVES

Evaluate the relationship between marketing and design
Define the term marketing
Explain who the customer is
Evaluate the meaning of the quality service culture
Describe the process of marketing planning
Evaluate various tools and methods used to ensure conformance to customers' needs and wants
Evaluate the benefits of customer-oriented marketing by ensuring fitness for purpose of goods supplied
Explain the basis and need for traceability
Describe the quality function development process
Evaluate control documentation used to effect traceability
Evaluate the effect of standards on traceability
Describe some of the perceived problems of traceability

CHAPTER OUTLINE

6.1 Introduction
6.2 Marketing and design
6.3 Who is the customer?
6.4 The quality service culture — providing customer service
    6.4.1 *Quality service effectiveness*
    6.4.2 *Managing the quality of service offered*
    6.4.3 *Benefits of the application of quality of service*
6.5 Marketing planning
    6.5.1 *The marketing mix*
6.6 Some tools and methods used to ensure conformance to customers' needs and wants
6.7 Benefits of customer-oriented marketing by ensuring fitness for purpose of goods supplied

6.8 Traceability
    6.8.1 *Control documentation used to effect traceability*
    6.8.2 *The effect of standards on traceability*
    6.8.3 *Some of the perceived problems of traceability*
6.9 Quality function deployment
    6.9.1 *QFD mechanics*
6.10 Chapter review
6.11 Chapter questions

6.1 Introduction

Designing a high-tech product used to take many years, e.g. motor vehicles used to take up to seven years to develop from conception through to first production model. Now, even with high-tech and very complex products, the design and first production unit coming off the production line can be between one and three years. Design affects how soon the customers get what they want, and what technologically they can have.

Design, as a process, has been severely confined as a management strategy for delivering products that conform to customers' specifications. Now that the quality revolution is here, the design element of quality management provides a major influence on ensuring and achieving quality products/services that sell. Design of products/services, and the processes that contribute to their enhancement, will be more constructive in the long run, in terms of costs to the organization and value for money to the customer.

Designing for quality is now seen as a competitive weapon. Designing what the customer wants provides opportunities for design engineers and marketers to enter into the quality management arena and take their rightful place.

6.2 Marketing and design

The integrated relationship between marketing and design is now seen as paramount in the quest to develop products that conform to customers' specifications and requirements. Marketing provides the data which designers use to develop products. Without a close relationship, the ineffective design of a product/service results, and a customer would move to competitors' products/services in order to gain satisfaction.

'Fitness for purpose' is a concept first described by Deming (1986). It has similar connotations to that given by Juran's (1974) 'Fitness for use' and Crosby's (1979) 'Conformance to requirements'. Peter Drucker in his book *Managing in Turbulent Times*, tells us that the survivors, public and private, in today's competitive environment will distinguish themselves in customers' minds either through clear product superiority, exceptional service or both. Providing products or services that continuously meet customer needs will require that the whole organization becomes quality-oriented and consequently customer-oriented.

Quality can no longer be seen as just meeting customer requirements. It is a continuous search for *added value* for customer requirements. This requires careful marketing to ensure that those needs, stated or implied, are provided as a meaningful basis for the design of the product or service. Organizational efforts must be directed towards what the customer expects, not what the organization *thinks* the customer expects.

Except as a source of customer requirements, quality is almost never mentioned in the literature, or discussed at quality conferences — few organizations seem to be actively pursuing quality in their marketing functions. This apparent omission of marketing from the commitment to quality does not seem appropriate. Marketing is usually the customer's first contact with the organization, and therefore, marketing is responsible for selling not only the product or service, but the organization as well.

In a product-driven organization, engineers and designers have often taken the position that 'our products are so good, they sell themselves'. They fail to recognize the requirements for effective marketing, which must be customer based. Marketing plays a vital role in every organization and the potential for improvement appears to be significant.

So, how can we briefly define marketing?

> Marketing is providing a product or service to satisfy customer needs, at a
> profit. Marketing is the social and managerial process by which
> individuals and groups obtain what they need and want through creating
> and exchanging products with others. (Kotler, 1991)

In a quality-oriented organization, marketing is more than the selling function. It is about the provision of meaningful data from which to develop products and services, and supplying those products and services that continuously satisfy customers' requirements.

In the quality marketing-oriented organization, the customer is the focus; whereas in the manufacturing-centred organization, they consider that the internal product is the focus and customers will follow.

6.3 Who is the customer?

To be effective, marketing — and what it stands for — must be developed internally throughout the organization. The development of the Juran concept — that everyone within the organization has a customer and a supplier, because everyone controls part or all of a process that itself has a customer and a supplier

— is fundamental to extending the marketing role internally. Since marketing affects every customer, then marketing affects every single person who works in an organization, whether they actually come into contact with the end customer or not. To generate a customer-first ethos requires the internal, as well as external, development of customer orientation.

Who, then is a customer? Everyone? Every individual is a customer within a defined process, the end result of which will eventually affect the quality features, both real and imaginary, of the final product or service. The 'secret' is to determine effectively, at each stage in the process, the needs of the customer following, then translate those needs into specifications that can be achieved, and monitor the process to ensure that the defined customer needs are being met. This should be a cyclical development.

To determine who the customers are, a process of market segmentation must be implemented. Here, a market is 'merely an aggregation of customers sharing similar needs and wants' (Kotler, 1991).

Segmentation is the process of classifying customers into groups with similar needs, characteristics and behaviours. To effect successful segmentation, marketers need to study and act upon, geographic, demographic, psychographic and other criteria that characterize the segment. Where the internal customers are concerned — because these are controlled directly by the organization and are generally a function of the structure, process and technology used — segments could change as the business process is refined. More importantly, because of direct control by the organization, process customers can be identified more easily and therefore the specification developed from their needs and wants can be more accurate. This would also allow the determination of job descriptions and person specifications that are more relevant to customer requirements rather than obsolete job descriptions that reflected what the organization expected they should do. This cannot be said about external customers. The division of a market into segments and targeting them — evaluating each segment's attractiveness — and selecting one or more favourable segments to enter is difficult to achieve effectively. Targeting also helps the internal customer/supplier relationship by ensuring that customer requirements are determined within the behavioural job aspects of the supplier.

Segmenting is a case of deciding which customers they are going to satisfy or are capable of satisfying. It is a balance between what the organization wants to achieve and the limitations of the present operational conditions. To further provide a differentiating means, marketers would determine a market position that provided a clear and distinctive place relative to competing products. Satisfying external customers is the primary aim, but to ensure this, the organization needs to organize its marketing orientation, both internal and external, to provide the means for this to occur.

When assessing customers, cognisance should be taken of what they need and require and:

1 The measurement of the quality of the output that provides the customers' supply.
2 The impact of any non-conformity.

Customers themselves should actively seek improvement of any supplied product or service. This moves very neatly onto the meaning of quality service.

6.4 The quality service culture — providing customer service

Much of the emphasis of service quality has been brought about specifically due to heightened competition, reducing the manufacturing base, and increasing the type, number and level of service industry providers. As such, most of service quality research has been on ways of classifying, conceptualizing and measuring that service quality (Hampton, 1993).

What therefore is customer service? Since all products have an element of service attached to them (in the least warranties, sales and after-sales service, etc.), then a service orientation is a necessary and vital area to develop. In the quality-oriented organization, customer service plays an important part for both internal and external customers.

Quality service therefore is not just limited to *pure* service organizations, it is a concept that must be developed and applied consistently in manufacturing-focused organizations also (O'Hara and Frodey, 1993). Yet, producing consistent quality services has not received as much attention as in the manufacturing sector (Mefford, 1993). Williams and Zigli (1987) indicate that 'progress is being made in defining service and service quality parameters, but imprecision and manufacturing mentality make the task difficult'.

Developing a *customer service culture* creates situations that reflect different departments owning the quality service problems that arise, when customer service complaints are made. This counteracts, for example, 'this isn't my responsibility — let the customer service people handle it', which is all too frequently heard in organizations today. It is further suggested that one of the main problems associated with poor customer service is that departments often work in isolation from each other, do not realize the knock-on effect of their activities and too often measure their performance against internal departmental yardsticks only — an internally oriented culture. The development of a positive customer service culture can therefore provide an effective mechanism for more productive interdepartmental relationships as well as consistent and effective quality of customer service. Understanding the focus and content of a customer service culture means changing quality-oriented behaviours within the organization. In viewing a service as a product, Albrecht and Zemke (1985) discussed several service characteristics. These include that services can only be produced at point of consumption, cannot be stored, are intangible and a sample cannot be sent for inspection.

These defined service characteristics affect how the service can be delivered, and knowledge of this leads not only to an understanding of the opportunities, but also of its critical limitations. Denburg and Kleiner (1993) take these various characteristics further and have determined that 'judgements on service quality are a function of customer expectations, which can be categorised into 5 overall dimensions'. These are reliability, tangibles, responsiveness, assurance and empathy. Thus, the service aspect takes on a more subjective approach and becomes much harder to manage effectively. However, service strategy must be reconciled with service expectancy, and therefore the strategy must change if there is a conflict.

6.4.1 *Quality service effectiveness*

A number of other authors such as Bowen and Schneider (1988) and Chase (1978) have attempted to define the attributes of services that complement Albrecht and Zemke (1985). These service attributes include the following:

1. They tend to be intangible in nature — The tangible product will always contain some aspect of service or intangibility, as it is the client who in the end will buy the product, and always requires some form or aspect of service. However, it seems relevant to assume that in the service industry it would be possible to derive an almost totally intangible product, e.g. insurance, or psychological advice.
2. They tend to be produced and consumed simultaneously — No intermediate distribution links the production of the service with its customer. Quality is created during the service encounter between service provider and service accepter — the customer.
3. The customer tends to participate in the production and delivery of the services consumed.

Chase and Tansik (1983) indicate that employees with high customer contact have a three-way interaction between 'themselves, customers, and the production process or technology'. On the other hand, low customer-contact personnel are involved in a two-way interaction between themselves and the technology. This provides some understanding of the problems and issues prevalent in managing service delivery. In the high-contact situation, there is a need to manage the customer and the complex server–customer relationship. The consequent training requirements are much higher in this situation because of the need to develop the interactional skills to effect appropriate levels of employee performance.

Lovelock (1981) discusses what he called 'service trinity', where the high contact service employee, at least in the mind of the customer, first, runs the organization, second, sells the output of the organization, and third, is equated by the customer with the organization's output.

In Lovelock's conception, the employee must demonstrate both managerial and marketing capabilities in addition to the technical skills (see Katz, 1974). The training of front-line staff becomes more important when they have to deal with individuals who they have not dealt with — as a company or with them as an industry — previously. Scripting is often used by for example, McDonalds, to ensure the consistent delivery of the service message to the customer. These findings have clear implications for the delivery of a quality service to customers. Complainers are a valuable asset to an organization. It is after it gets the chance to put things right and fails, that genuine and loyal customers leave for the competition.

6.4.2 *Managing the quality of service offered*

Nel and Pitt (1993) suggest that if 'service quality is to be a source of competitive advantage, and differentiation, it must be carefully managed'. Snee (1993) indicates that organizations which concentrate on learning and improving their internal processes rapidly improve product and service quality. Consequently, managing internal processes means ensuring that customers become satisfied and that the organization is able and willing to respond to changing customer needs. Robust design of internal processes provides a basis for this to occur.

Features that organizations delivering outstanding service have in common include:

1. *A well-conceived and implemented strategy for consistently developing and delivering quality service.* The quality-oriented organizations have discovered, invented or evolved an all-encompassing, long-term service strategy of what they want to achieve. This service strategy directs the attention of the people in the organizations toward the real priorities of the customer.
2. *Customer-oriented staff — both internally and externally.* Staff who deliver quality service have been encouraged and rewarded in order to direct their attention continuously on the identified needs of the customer. This leads to a level of responsiveness, attentiveness and willingness to help that marks the service as superior in the customer's mind.
3. *Customer-friendly systems.* The delivery system that backs up the service is designed for the convenience of delivering the service to the customer, rather than for the convenience of the organization. The physical facilities, policies, procedures, methods and communication processes and staff, all say 'customer service'.

This has important implications, not only for the manufacturer, but also — and perhaps more importantly — for the level and type of service offered to customers in a service setting.

Thus an external customer orientation cannot be achieved without an internal focus, set against staff involvement and commitment to delivering quality service. It is this philosophy that reinforces the overall quality culture of the organization, as the icons of quality service are the important characteristics that customers see, perceive and react to. The suggestion of just applying a manufacturing model of quality without special adaptions for service orientations could be construed as a major pitfall. Further, taking too long to make the quality service change could be just as damaging as not attempting to carry out the service change in the first place.

Again, Albrecht and Zemke (1985) suggest that there are several factors that make it difficult to define and improve the quality of service developed and delivered effectively. Services, because of their intangible nature, are seen as less easy to break down and discuss in precise, operational terms. As a consequence, management and staff often tend to talk about service delivery in vague generalities. Since the nature of the service cannot be absolutely defined, and not altogether consistent or standardized, it lends itself to actions that are variable, inconsistent and difficult to control in the delivery of that service. This brings about the experience by internal or external customers of varied levels of service, on different occasions when interacting with the same or different employees. A shocking discovery is that front-line staff (who provide the service most often to external customers) have the lowest skills, the lowest remuneration and are the least motivated members of the service organization. Upskilling therefore, benefits the organization, and the service interface, resulting in more economical and effective service delivery.

Service can be explained in terms of a continuous cultural awareness of what it takes to satisfy a customer and do it each time, without fail. In light of this, the application of quality management to services essentially means dealing with the culture. This requires the measuring and evaluation of the:

1   Baseline or current frequency of the service output behaviour.
2   Motivator required to provide the service behaviour.
3   Improved quality and/or quantity of service delivery.

These behaviourally reinforced settings are best applied in — though certainly not limited to — standardized service operations, in which customer needs can be predicted in advance; but are less applicable to professional services, such as management consulting or teaching or training situations.

6.4.3 *Benefits of the application of quality of service*

1   *Consistency* — Customers can rely on the service every time. This means not only on the delivered product, but also on the increasingly developed expectancies that go beyond that. For example, customers tend to become accustomed to the delivered quality of service, and next time may expect more. Consistent, therefore means, not only delivering now what is expected, but also in the future.
2   *Focus* — The quality-oriented organization is paradoxically focused on itself, its processes and direction, whilst simultaneously ensuring that changing

3 *Knowledge* — Through the application of service quality, the quality-oriented organization is being taught about its capability, and reinforcing its operational commitment towards heterogeneity, speed and competence of the service offered.
4 *Training* — Management consider that in order to deliver effective service, their staff must be trained appropriately in order to project the quality message.
5 *Teamwork* — The application of quality service assists every individual in the organization to help everyone else. As a consequence, a more outward-oriented approach is developed that forms the basis for integrated teamwork.
6 *Control* — The quality-oriented organization would control its processes more effectively and therefore result in less wastage, fewer communication problems and increased internal work satisfaction.

6.5 Marketing planning

Marketing assists the strategic planning process. This process can conceivably be effectively implemented, only if it is based on an effective marketing strategy, while using a well-balanced marketing mix.

The marketing plan will be developed as one of the functional plans, which should stress the customers' needs and requirements. In a quality and customer-oriented organization, the marketing plan drives the marketing function as the integrative function within the whole organization. The focus of the marketing management process is the marketing plan, which reflects the determination of the marketing mix strategies. Because of the need for marketing to be an integrative function, the marketing plan should provide ready support of the objectives and strategies of all other functional plans by providing necessary marketing data when needed.

6.5.1 *The marketing mix*

This is defined as 'the set of controllable marketing variables that the firm blends to produce the response it wants in the target market' (Kotler, 1991). The marketing mix (first determined by McCarthy, 1961) provides the framework which consists of everything an organization can do to influence the demand for its product and is seen as follows:

- *Product* — The product/service offered and exchanged in the market.
- *Price* — The actual cost customers must pay for the product/service.
- *Place* — The organization's ability to make the product/service available to the customer.

- *Promotion* — The communication activities that strive to motivate customers to purchase the organization's product/services.

The above are determined as the main constituent areas within marketing. All elements of the marketing mix must work together to eventually say 'quality'. The marketing mix strategies are a function of the internal arrangements that the organization wishes to use as the basis for determining the external marketing orientation. An effective marketing programme blends all the marketing mix elements into a co-ordinated programme designed to achieve the organization's marketing objectives, which are determined from close study of what its customers want. This in turn provides the basis for design actions targeted to effective manufacturing and process elements (Fraser-Robinson, 1991).

Fraser-Robinson (1991) indicates that since the 1960s the general movement from mass to segment to niche to customized marketing can be seen to be developed. What this really means in practical terms is that effective marketing mix strategies must rely heavily on market research and marketing intelligence. The difficulty comes not so much from a problem of the development of the proposed marketing strategies, but from the effective application of the necessary management skills and competencies needed to implement the marketing strategies so developed.

6.6 Some tools and methods used to ensure conformance to customers' needs and wants

Methods that can be included in the development of product design are:

1. Quality Function Deployment (QFD) (Eureka and Ryan, 1988).
2. House of quality (Hauser and Clausing, 1988).
3. Taguchi methods (Taguchi and Clausing, 1990).

The basis for the use of these very important methods is the marketing research carried out to provide the base data. To determine appropriate customer needs, market research should be carried out. Garvin (1987) suggests that 'Shoddy market research often results in neglect of quality dimensions that *are* critical to consumers'. Garvin's quality dimensions can be used to good effect both inside and outside an organization. These are discussed in greater detail in Chapter 4.

The teaching of market research, which gives marketing skills to workers, ensures that the customer quality chain by Juran (1974) receives the best attention to detail. This can only ensure that, once the product or service has been designed, the specifications are adhered to throughout each customer quality chain (internal), and culminate in a satisfied customer (external).

6.7 Benefits of customer-oriented marketing by ensuring fitness for purpose of goods supplied

Marketing has a major role to play in ensuring *fitness for purpose* or *fitness for use* of goods supplied. The basis for this role is the use of marketing research techniques that deliberately try to determine the customers' needs and wants and translates them into meaningful specifications that are readily understood and acted upon by all concerned with ensuring the quality of conformance to those developed specifications. This practice underlies the development of techniques such as QFD that provide the many benefits which can only be said to be effective when the customers' voice is heard right through the development design process, the manufacturing process and through to delivery.

The development of an integrative marketing function would ensure that all planning activities are related to developing real and effective customer specifications throughout the customer chain. Effective monitoring of output from each supplier will ensure conformance to the designed customer specifications and requirements. Long-term customer relationships would develop, even if customer specifications change because of the continuous assessment of customer needs, both internal and external. This would be through customer-oriented planning to provide the strategic orientation necessary to develop the consistency of quality action. Competitive pressures, much more discriminating customers (as far as parting with their money) and recessive market structures, require increased market segmentation and the forcing of a lowering customer base. Fitness for purpose or use thus becomes the major orientation for survival and increased competitiveness.

6.8 Traceability

Traceability of design requirements from customer/market research through to supply, requires continuous information interactions, where later design outputs are evaluated against previously designed specifications. Essentially, this becomes hierarchical in nature, where higher level specifications direct the more detailed design outcomes. However, data transference through the hierarchy requires the effective crossing of the various interfaces deemed to exist through the customer–design–supply process. To illustrate this, consider Figure 6.1.

Between each section of the product development, there is an interface that needs to be crossed in order for the design specifications (on the left), to become the design output (on the right).

To provide an inherent means of traceability through customer/supply, design methods that can document each stage of the design process need to be used. Methods that do exactly this are few and far between though, because of

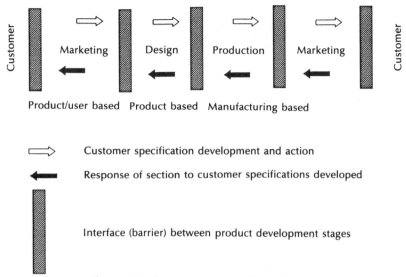

Figure 6.1 A customer–supply model.

the complexity of the design projects and because the cultural definition of the donor organization has not changed to take advantage of the method used effectively.

Design is separate from the manufacturing process, as 'befits' a Tayloristic attitude (Fraser-Robinson, 1991) with separation of the planning from doing. In this case, design takes on the responsibility of developing both product and process requirements, not only of the design project but also of the manufacturing process. If, as the Tayloristic attitude requires, design is carried out in isolation, manufacturing will respond with the need for increasing design changes that gives extra rework, higher production costs and a higher overall unit cost for a given product. Therefore methods that generate multiple quality perspectives, through teamwork design, benefit greatly from lower design costs, lower change charges and greater ability to target — right first time — customer needs and wants.

6.8.1 *Control documentation used to effect traceability*

No matter what methods are used to effect product development, there is an inherent need to determine how traceability is effected. Within each method, various means are used to direct the customers' voice.

In requirements planning, defining functional activities requires information about events and control, and for each function there is a full description of an activity. Each activity is traceable back to a viewpoint of the whole product, inclusive of its environment and the viewpoint is a derivation of the design brief between customer and designer. Thus this provides a means of building up a definable path, so that interpretation of the design brief can be examined for compliance with the standards or original intentions.

A key process is understanding the perceived implementation by the customer, translating that into a functional description and implementing those functions within real-world constraints. Quality of design can only be demonstrated if the above is traceable.

Methods of traceability are determined as:

1   *Embedded traceability* — where every statement, diagram element or drawing is referenced for its response to and consistency with other requirements.
2   *Review traceability* — where a design review process is actioned to allow the consistency and accountability to be demonstrated by key individuals in the design team.

The mechanism normally adopted to ensure traceability is design change control. This is done to trace a design development, establish the baselines, gain agreement on design change and allow all involved to understand the status of all that is being controlled.

6.8.2 *The effect of standards on traceability*

In a commercial environment, in Europe at least, there is an increasing requirement of vendors to adopt the BS EN ISO 9000 (BS 5750) Quality Management System standard. This means that designs *per se* will need to subscribe to this system in order to meet the standard.

Consequently, traceability will need to become the norm, rather than the exception. This will provide legal defences, in the case of contract disputes and negligence claims for product liability, e.g. in the food industry. Gone are the days when a new product can be trialled on the general public. Consequently, traceability will take on a new air of relevance than previously accepted. Models are available that seek to provide working examples of the design quality documentation relative to BS EN ISO 9000.

This indicates the complexity and scope in which traceability must be enacted and provides an enormous problem related to determining exactly the cause of a design output and where and how they relate to each other. Consequently, methods have been adopted to assist with this process and the use of relational databases becomes a must, not a luxury, in these circumstances. This is especially so, since some projects may have many hundreds of thousands of tasks that need to be traced to the defined customer requirements and needs.

This brings us to the point of who is effectively capable of assessing the quality system against the standard when complex engineering designs are being developed and who would have the objectiveness to facilitate the effective assessment on the many thousands of procedures and work instructions. The problem of traceability rears its head, not just for the customers' voice, but also in the process of ensuring that the customers' voice is heard appropriately throughout the manufacturing process and through to delivery.

6.8.3 *Some of the perceived problems of traceability*

Traceability from customer through to supply is prescriptively achieved through the use of various methodologies, such as QFD. In practice, cultural, political and the quality views determined above all play a part in resisting the voice of customer percolating through the organization.

It would seem quite evident that because the quality 'views' are quite deep, then communication of the real meaning of the customer — who may have another quality view — is tinged with meanings attributed by each different quality view group within the organization. What results essentially is autopiotic systems (Garvin, 1987), which are powerful forces that try to shape 'reality' in their own image — self-referential systems. This effectively means that each quality view, including the customer, tries to influence the outcome of the design specifications and products in the shape of their own image, i.e. marketing using a product/user strategy would seek to influence other groups in the organization to accept their design orientation. Incidently, during the 1980s marketing, *per se*, was largely successful with this practice, but has still failed to implement its quality view effectively, because of the strong power of internal quality views, such as manufacturing.

Traceability within each group is reasonably successful, relative to their system of accountability, but across barriers, which essentially is the traceability requirement, it falters.

6.9 Quality function deployment

Quality function deployment (QFD) is a formal design methodology that provides an holistic approach to effective product design. This method is essentially accepted as a rational design method, rather than the more subjective creative methods commonly found in marketing/engineering/production teams.

At Toyota Motor Corp (Eureka and Ryan, 1988), QFD reduced the number of engineering changes during product development by 50 per cent. Development time and start-up cycles were also cut, reducing the overall development time to market — a major form of competition today. The one major activity which allowed this to occur was good documentation procedures and the demonstration of traceability of quality from the customers' needs to delivery and through to servicing.

QFD is a methodology for the effective planning and development of quality goals that serve to meet the perceived customers' notion of fitness for use. The term QFD is a bit misleading though — it is not a quality tool, but a visual planning

QUALITY OF DESIGN

tool. It is a systematic way of ensuring that the development of product features, characteristics and specifications, as well as the selection and development of process equipment, methods and tools, are based on the developed demands of the customer.

Essentially, QFD is a means by which the voice of the customer is effectively translated into the product/service that satisfies them continuously. Besides all this, QFD relies heavily on the marketing process and more particularly, market research. It makes it easier to sell products to customers — the right customers — every time. With QFD, broad product development objectives are broken down into specific, actionable tasks.

Team effort provides the integrative approach necessary for effective consensus and the development of future committed planning actions. The process is achieved via the use of a series of matrices and charts that deploy customer requirements and related technical requirements from product planning and design to process planning and manufacturing.

QFD is essentially divided into two parts. Part one identifies the customers' product/service needs and wants and translates these into designated specifications.

Part two develops the internal organizational requirements necessary to satisfy completely the identified customers' needs and wants. The heart of the technique of QFD is the application of what has become known as the *House of Quality* (Hauser and Clausing, 1988) — matrices in which the HOWs of one stage are related to the WHATs of the following stage (see Figure 6.2).

This provides a hierarchical determination from designed customer specification to work specifications and arrangements. The outcomes of QFD are

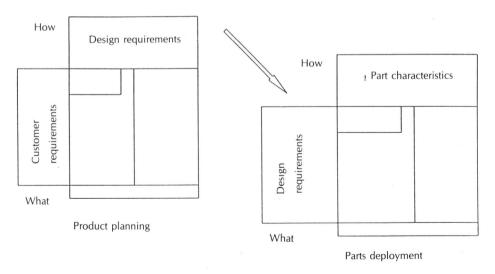

Figure 6.2 Quality function deployment — 1.

the design of more competitive products, in less time, at lower cost and higher quality (in the customers' viewpoint). QFD users report increased customer satisfaction, market share, improved designs and reduced warranty claims.

The first QFD matrix serves as the basis for any and all succeeding QFD phases. The information provided in this initial phase is used to identify specific design requirements that must be achieved in order to satisfy customer requirements. Therefore the need to get this right first time is paramount, as it considerably affects all following quality capabilities. Design specifications are the output from this stage.

6.9.1 *QFD mechanics*

The basic process begins with the development of the notion of customer requirements, which are usually loosely stated qualitative characteristics, such as 'looks and feels good', 'easy to use', 'safe', etc.; characteristics determined in the language of the customer (so that the voice of the customer is heard) and therefore thought to be important to the customer. This customer's voice often defies objective quantification and as such becomes difficult to effect a precisely structured design approach. This phase is extremely important in regards to traceability, because if the interpretation is incorrect, then traceability back to the customer will never occur.

During product development, customer requirements are translated into internal specifications, sometimes called design requirements. These are generally global product characteristics (measurable) that will satisfy customer requirements, if effectively executed.

QFD can be thought of as a design method with a four-part process:

1   Part one — product planning.
2   Part two — product design.
3   Part three — production and service process planning.
4   Part four — workplace activities and tasks.

At the heart of the first phase of QFD is the House of Quality, which is determined to contain a relationship matrix so that product planning can be used effectively to describe, quantify and qualify customer needs and requirements, design requirements and competitive analysis (see Figure 6.3).

Today, the best quality-oriented organizations deploy the voice of the customer to help determine important product attributes. Using this type of product development effort, these organizations focus on product/process planning and problem prevention issues.

QFD is one of the methodologies used to make the transition from reaction to preventative — from downstream, manufacturing-oriented quality control to upstream, product-design-oriented quality control. However, this method is an amalgam of at least seven smaller methods. These are the affinity diagram, interrelationship diagraph, tree diagram, matrix diagram, matrix data analysis,

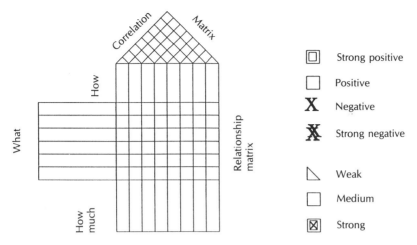

Figure 6.3 Quality function deployment – 2.

process-decision programme chart and the arrow diagram (Taguchi and Clausing, 1990) — the new seven tools of quality.

6.10 Chapter review

Now that the quality revolution is here, the design element of quality management provides a major influence to ensure and achieve quality products/services that sell. Designing for quality is now seen as a very useful competitive weapon. Marketing is the process of supplying exactly what satisfies customers. Designing what the customer wants provides opportunities for design engineers and marketers to enter into the quality management arena and take their rightful place. Thus, the integrated relationship between marketing and design is now seen as paramount in the quest to develop products and services that conform to customers' specifications.

To determine who the customers are, a process of segmentation must be implemented, where customers are classified into groups with similar needs, characteristics and behaviours. To effect successful segmentation, marketers need to study and act upon geographic, demographic, psychographic and other criteria that characterize the segment.

Quality of service is seen to be as important for manufacturing as it is for service industries. Services tend to be more intangible in nature, to be produced and consumed simultaneously and the customer tends to participate in the production and delivery of the services that they consume. Managing internal processes means ensuring that customers become satisfied and that the organization is able and willing to respond to changing customer needs.

Features of organizations delivering outstanding service include a well-developed and implemented strategy for consistently developing and delivering quality service, a customer-oriented staff and also customer-friendly systems. Benefits of quality service include, consistency, focus, knowledge, training, teamwork and control of the operation.

Marketing planning consists of building the organizational mission, the development of the business portfolio and the functional plans. Out of this, the marketing plan is constructed.

The marketing mix is defined as the set of controllable marketing variables that an organization produces to create the response from its various target markets. All elements of the marketing mix must blend together effectively to say 'quality'.

Market research is becoming a very important tool in managing quality. It is the method in which data are developed, first about what customers want and need, and then second about monitoring how the organization performs in satisfying those customers.

There are many influences that determine the effective traceability of design specifications and design products within an organization. These influences include the customer and various groups within the organization who have a substantial input into the effective design, development and production of products that aim to satisfy customer needs.

Methodologies that create 'teamwork', e.g. QFD, try and eradicate problems of self-referential systems. This still fails, and the use of simultaneous engineering techniques tries to allow for individual group requirements and also ensure that the customer gets what they want. Political strategies abound which seek to ensure the survival of these autopiotic systems and these in themselves provide much of the resistance to changing quality perspectives.

Traceability is an ideal that in practice is only effected within limits. Assumptions are made about what design specifications mean and what design products should be derived. However, the adoption of international standards like BS EN ISO 9000 will mean the generation of commitment to traceability and failure to demonstrate this will effectively nullify many contractual positions and lead to organizational demise. This is the new face of quality. Without the standard, many engineering projects will be out of reach and the standard brings with it more 'bureaucracy' and management control; it is this control that may be a key resistance issue within design, because of the paradox of control versus creativity.

6.11 Chapter questions

1. Define the term marketing.
2. Evaluate the relationship between marketing and design.
3. Explain who the customer is.
4. Describe the process of marketing planning.
5. Evaluate the benefits of customer-oriented marketing by ensuring fitness for purpose of goods supplied.
6. Explain the basis and need for traceability in design.
7. Describe and evaluate the house of quality as a design method.
8. Evaluate the place of market research in quality management.

CHAPTER 7

Organizational structure and design

CHAPTER OBJECTIVES

Define organizing and organizational structure
Evaluate organizational design
Discuss various job design methods
Compare and contrast centralization versus decentralization
Evaluate implications for organizational effectiveness

CHAPTER OUTLINE

7.1 Organizing
7.2 Organizational structure
7.3 Organizational design
7.4 Job design
    7.4.1 *Job design methods*
7.5 Centralization versus decentralization
7.6 Implications for organizational effectiveness
7.7 Chapter review
7.8 Chapter questions

7.1 Organizing

Effective organizing means the development of structuring tasks, processes and resources so that organizational objectives are met efficiently. Organizing for quality means ensuring that the organization meets the quality objectives it has set. It also means providing flexibility in action and outlook to respond to changing internal and external environments. Internally, it means providing a flexible and more responsive approach to managing human resources and the technology and the processes they use. Externally, it means providing a more flexible and responsive approach to managing customers' needs and wants, and to manage the continuing pressures of the competition. This undoubtedly means structuring an organization to provide for this. The structure can inhibit or enhance flexibility and/or affect how people are able to use the available technology. The way an organization structures itself has dramatic affects on how it manages quality.

7.2 Organizational structure

Structure means the development of a clear framework of tasks and responsibilities that contribute effectively to the efficient operation of an organization by directing the behaviour of individuals, groups and departments to effect set organizational objectives. Structure is seen in all organizations, and can mean hierarchy, but some structures are less effective in given situations than others.

Today, structures that are seemingly flat provide greater flexibility and speed of response than 'taller' hierarchical structures, which are slower to react and change. Satisfying customers can be achieved with either, but what happens when the outside environment forces changes that affect the customer's expectations on the type of quality products or services? When customers change, so must the organizational response — and this is where flexibility and the learning organization come in. But flexibility means the provision of less of a structure to inhibit, less bureaucracy and less control. How then, can an organization demonstrate its commitment to a customer through quality if the quality system is not there? This means there will inevitably be some form of structure, some element of bureaucracy, some application of control. The challenge then is to balance the need for control with the need for flexibility.

Organizational structure reflects past management decisions. It also reflects the management response as to who will do what and how (in the sense of directed responsibilities, rather than the actual methods used). The structure must also represent the quality strategy that management is adopting. Hierarchical structures will need to be flattened in order to provide a visible delineation of management's quality intentions. Organizations owe much of their inertia and resistance to change to their structure, and it is this that creates problems when management changes its quality strategies. Structure must come first, and must be seen to be appropriate before the culture of an organization can change. Changing structure therefore has a profound effect on how an organization can change to become quality-oriented. Structure therefore provides the *bones* on which the *flesh* of quality rests.

An effective structure for one organization may not be effectively applied in another. Each organization is unique; it has a singular set of internal and external environmental opportunities and constraints. Many worker-related problems can be attributed to an incorrectly developed structure. The challenge is therefore to ensure that an organization has the most appropriate structure in place and working effectively to meet those internal and external problems and opportunities.

Problems of structure include:

1. Organizational subunits (departments or groups) becoming independent, developing their own methods of operation and their own quality objectives to satisfy. This affects the coherent direction of an organization as conflicts abound and politics, rather than good managerial practice, provides the basic management tool.
2. Structure determining strategy, rather than the reverse. Management cannot allow the structure of an organization to dictate its overall or specific quality strategies. This is a recipe for failure, as the strategies that can be applied are relative to the *old way of doing things* (considered as structured inertia) and limit the organizational ability to make effective changes in light of the circumstances facing it.
3. Task and individual responsibilities being ambiguous. These create difficulty in relation to skills development, applying appropriate rewards, evaluating performance and optimizing process times.

7.3 Organizational design

Determining the operating relationships between the amount of specialization of a job, the amount of delegation and authority given to the incumbents of a given job and the span of control in which individuals have to operate is the application of organizational design. Managers making choices about the degree of application of these factors must bear in mind that a quality-oriented organization would tend towards a high degree of delegation, authority and span of control, coupled with a reduced amount of specialization. Child (1977) indicated that the development of organizational structure should consider the following factors:

1. The assignment of tasks and responsibilities in order to define and prescribe job roles clearly.
2. The conscious development of segmenting the organization into specific units that are in themselves self-actioning.
3. The development of hierarchical requirements in order to expedite communication, command and decision operations, delegation and co-ordination efforts.

An organizational chart is a visible outcome of this development. It projects, both internally and externally, the relationships described above. It also denotes line (command) and staff (support) details that provide the fabric for the conscious and unconscious operation of the organization. The organizational chart also indicates the chain of command — scalar chain — where individuals know the clear line of authority. This principle is seen in many quality circles, as they anticipate that the leader position will be chaired by different people at different times — the job role and responsibilities stemming from it do not change. The organizational chart also has limitations, not of what is visible, but of what occurs in practice in the organization. These include:

1. The chart depicts what should be done, it is therefore prescriptive.
2. The chart does not indicate how jobs are broken up, or what methods, procedures or processes are used.
3. The chart does not indicate the effective competency of the individuals holding each job position depicted.
4. The chart does not indicate changes to the hierarchy or changes to horizontal segments of the organization.
5. The chart does not show how individuals communicate, both vertically and horizontally.

Models of organizational design include:

1. *Mechanistic* — Rigid structures that use methods to achieve efficiency, by grouping specialist functions together and through the application of strict guidelines such as rules and procedures develop a centralized orientation. This typifies Weber's bureaucracy, where authority rests with organizational position. Communication is predominantly vertically oriented.
2. *Organistic* — Flexible structures that are inherently innovative, with much less bureaucratic hinderance placing greater emphasis on individualization and communication both horizontally and vertically. It is more human in orientation than mechanistic structures.

Burns and Stalker (1961) researched these organizational designs. Many of today's organizations — including government agencies — are moving towards organistic structuring. As a development, matrix organizations — project oriented — have emerged in the last thirty years. These structures are considered short term and temporary, they provide flat cross-functional organizational operations (Ford and Randolph, 1992). What is important in organizations is to view the predominant form as the totality of its functional orientation.

Most organizations exhibit the three models indicated above at some time or another. However, it is the dominant form that endures and inevitably directs the survival of the organization. In the quality revolution, it is expected that the organistic form will be pursued and become dominant.

Another structure that is developing because of increasing use of computer technology is the contract or network structure, where a centralized operation (predominantly small in size) directs its requirements through delegating work to other organizations who perform manufacturing, marketing or other related business function. Much of this form was carried out in the past by marketing or production specialists who worked on a small element of an organization's workflow that would otherwise cost the organization greatly in terms of speed to market, learning curve and delay to operational programmes. This structure relies heavily on effective quality relationships with the independent suppliers of services and products. There is also an increase in internationalization of these operations, where product development is carried out in the United Kingdom, manufacturing in China, and marketing and sales in the United States. There is a greater requirement for trust and mutual development and the effective management of these enterprises hinges not only on the reliability of supply, but on the consistent delivery of product/services to customers.

7.4 Job design

Job design provides the greatest influence on organizational structure. The larger an organization, the greater the likelihood of specialization. This occurs because there is a limit to the amount of work — physical or mental — that an individual can perform. Consequently, as an organization grows, so does the need and pressures for specialization. Taylor (1915) used specialization as the basis for his methods of working. Thus, differing work requires different skills, attitudes and aptitudes. Specialization therefore allows individuals to develop more focused skills and competencies in order to carry out their tasks and job responsibilities more effectively. Job design allows individuals to be more specifically targeted to jobs that are more relevant to their capabilities. This targeting ensures that changing organizational operations which better meet customer requirements will ensure that quality products/services are provided efficiently and effectively. Job design therefore is a very important aspect of the quality-oriented organization. It is not sufficient to give people extended job specifications — they must be trained and educated and they must be motivated sufficiently to carry out that job.

7.4.1 *Job design methods*

There are four job design methods as illustrated in Figure 7.1. These are:

1. *Job simplification* — the process of narrowing the specializing of a job, so that the job holder has fewer activities to perform. This is Taylorism. The focus is efficiency. It harbours boredom, loss of motivation, little mental stimulation and very little room for innovation on the part of the operator.

2. *Job enlargement* — the development of job specifications that increases the variety of activities that an individual carries out. It is more challenging and there is room for increased mental stimulation through increased innovation and risk-taking. It also provides improvements in operator skills and competencies through increased skills development, training and education. Particular issues arise when considering this approach of job design regarding

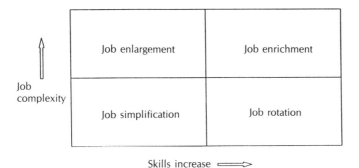

Figure 7.1 Job design.

unions, who provide much resistance to this type of change. Much of this resistance was exhibited in the United Kingdom with the introduction of quality circles. Japanization issues have affected demarcation lines of singular job specifications and union power has diminished as a consequence. However, just increasing similar job activities burdens individuals who are already bored and frustrated. All it does is increase job scope, not job depth.

3   *Job rotation* — the rotation of individuals through different job sets and a planned order. Job rotation increases skills and releases workers from the effects of boredom related to job simplification and job enlargement. It provides flexibility in workforce management and job assignments. It also provides greater motivation to individuals if the rotation takes the worker to different locations and allows greater meaningful contact with different workers. This approach can be applied to worker or management activities with equal results. Potential anomalies include treating the rotated worker as extra help and giving them only menial tasks, and the development of departmental or group loyalty can be difficult.

4   *Job enrichment* — the process of developing job contents that increase job skills, the potential for individual growth through education and training, achievement, recognition and responsibility. First described by Hertzberg (1966), it is matching Hertzberg's higher motivation levels. Essentially it provides control over job elements that the worker would otherwise not have. However, workers in quality environments that subscribe to quality management standards may actually be working to the reverse of this approach. Hackman and Oldham (1980) developed a job characteristics model which involved three major elements — core characteristics (skill variety, task identity, task significance, autonomy and performance feedback), psychological states (work is perceived as being meaningful) and job outcomes (responsibility for actual outcomes, and growth satisfaction through the work itself).

Job enrichment is considered the highest level — in terms of skills and sense of achievement given to staff — to achieve in job restructuring. However, it is also the most difficult to substantiate, because individuals learn and develop and over a period time what they considered motivational may become routine. This presents structuralists with major problems in today's organization — even if it attempts to become quality-oriented.

Another approach to these methods is the process of assessing structural alternatives. That is, rather than changing the actual jobs themselves, change the job schedule by providing alternative work programmes. It is based on the assumption that work and social lives intermix, and therefore it is reasonable to provide a mechanism that balances the competing associated needs. This is done through various programmes such as flexitime — a programme that specifies certain core hours of attendance, outside which the attendance time is flexible within a given timeframe. Flexitime reduces turnover, increases worker morale and indicates management's sensitivity to workplace problems. Another aspect is job share, where two or more individuals share a single full-time job. It is particularly useful when workers want to work part-time.

Co-ordination requirements are increased as an organization moves further down the quality track. They affect communication needs and the ability of staff to carry out the increasing need to become flexible in work output and operational performances. Factors influencing the degree of co-ordination in organizations include:

1   *The degree of formalization* — The depth of written quality policies, quality procedures and other documentation that determines how people behave in relation to job requirements and job actions. In organizations that have a formal quality management system, formalization is quite high — but the degree of formalization increases towards the lower levels of operation. This is rather unbalanced and illustrates that in these quality-oriented organizations people mostly do rather than create. Size has a bearing on formalization too. The smaller the organization, the less likely that it will be run in a highly formalized way, resulting in less bureaucracy but less guidance to its workers.

2   *Span of control* — The actual number of subordinates a manager controls in their everyday work situation. Too few workers can result in a critical overseeing of work-related activities, giving rise to increased supervision that may not be warranted, with loss of worker discretion. Too large a span can lead to manager work overload and inefficiencies and results in ineffective management (Child, 1977). Barkdull (1963) suggested that the span of control can become wider when there is:
    (a)   Low problem frequency — resulting in less managerial interaction.
    (b)   High managerial competence — allowing greater scope for numbers of subordinates controlled at any given time.

(c) High motivational elements of the actual work — opportunities for subordinate growth and skills development means less managerial interference.

(d) Additional support — resulting in greater managerial attention to the everyday issues facing the subordinate workforce.

3   *Authority* — The right to make decisions, control other individuals' work outputs, and direct the process of work relative to a given position's tasks, goals and requirements set by the organization. In a quality-oriented organization, authority can rest with the front-line or production worker to make decisions that used to be made by management. For example, a few years ago in the car industry, it was considered only a manager's job to stop the production line in the event of a quality problem. Today, many car manufacturing plants have established this as a right of the production-line worker.

4   *Delegation* — The assignment of part of a manager's job to a subordinate. In order for a quality-oriented organization to function effectively, a manager must be able to delegate. Problems exist with the implementation of this. How much should be delegated? When? Common problems include resistance to delegation by both management and workers fearing blame for failure, lack of time, lack of commitment to training and education or just plain laziness by managers.

Delegation can be defined generally in the following ways:
(a) Delegation is the pushing down of authority from superior to subordinate (Dessler, 1985).
(b) Delegation is the process by which managers assign tasks and the authority and responsibility to complete them (Mukhi *et al.*, 1988).

7.5 Centralization versus decentralization

Centralization refers to the degree of dispersion of authority in an organization. The trend today is towards decentralization, in order to give lower managers and workers more direct control of the environment in which they are expected to operate. It means using less line management, so fewer middle managers are needed. This in itself creates great resistance to the application of these types of measures. The quality-oriented organization is therefore flatter, with less middle management and a greater decentralized feature. Decentralized organizations exhibit faster reactions to competitive pressures and changing customer needs and wants, the development of managerial skills in decision making, and the development of measures that rest squarely on responsible autonomy.

7.6 Implications for organizational effectiveness

The organizational design has major implications for flexibility, innovation and entrepreneurship. In this regard, the structure denotes in the short term, the what and the how of an organizational response to competing pressures — both internal and external. Another major aspect is the effect of the stage of the organizational lifecycle. Here, each stage in the lifecycle denotes expressly what can be achieved in the short term.

When the effectiveness of an organization is evaluated, we examine the value of its responses to the organization current, as well as its potential, gains in market share, added value, wellbeing of customers' needs and that of suppliers and shareholders. Much of the organizational gains are due to its structural developments and orientation. Consequently, structure enhances or limits what an organization can do now, but does not limit what can be done in the future. That said, if the organizational culture prevents changes being made to the structure of the organization, then that in itself may drastically limit the future orientation of the organization. The quality-oriented organization is designed to be flexible in both application now and in future. However, the cost for this is the need for control, which is somewhat offset by the very nature of quality management practices — decisions being made at the point of control.

7.7 Chapter review

Organizing is defined as the development of structuring tasks, processes and resources, so that organizational objectives are met efficiently. Organizing for quality means ensuring that the organization meets the quality objectives it has set. The way an organization structures itself has dramatic affects on how it manages quality.

Organizational structure is defined as the development of a clear framework of tasks and responsibilities that contribute effectively to the efficient operation of an organization through directing the behaviour of individuals, groups and departments to effect set organizational objectives. Structure does not always mean hierarchy, but there seems to be some correlation between low levels of hierarchy and the quality-oriented organization. Therefore, the structure directly represents the quality strategy that top management has adopted. Problems of structure include the equitable balance between dependent and independent operation for subgroups and departments, structure dictating strategy rather than the reverse (structural inertia), and task and individual responsibilities becoming ambiguous in the face of mounting change.

Organizational design concerns the effective determination of operating relationships between the amount of specialization of a given job, the amount of delegation and authority given to the incumbent's given job or post and the span of control that individuals have to operate.

Job design provides the greatest influence on organizational structure and therefore efficiency. Four design methods are:

1 *Job simplification* — The process of narrowing the specializing of a job, so that the incumbent has fewer activities to perform — deskilling.
2 *Job enlargement* — The development of job specifications that increase the variety of activities that an individual carries out.
3 *Job rotation* — The planned rotation of individuals through different job sets.
4 *Job enrichment* — The process of developing job contents that increase job skills, and the potential for individual growth through education and training, achievement, recognition and responsibility.

Centralization versus decentralization issues arise out of the balance necessary to ensure an effective and efficient organizational design. These affect groups and departments and also affect the speed of reaction to competitive and customer pressures, the number of line managers required and the amount of responsible autonomy allowed in the organization.

7.8 Chapter questions

1 Define organizing and organizational structure.
2 Evaluate organizational design in the context of quality management.
3 Discuss the various job design methods.
4 Compare and contrast centralization versus decentralization.
5 Evaluate implications for organizational effectiveness.

CHAPTER 8

Leadership

CHAPTER OBJECTIVES

Define quality leadership
Evaluate how leaders influence
Discuss the various theories of quality leadership
Define the term motivation
Evaluate the nature of motivation
Discuss the various theories of motivation

CHAPTER OUTLINE

- **8.1** What is quality leadership?
- **8.2** Strategic quality leadership — the need
- **8.3** How leaders influence — the power stakes
- **8.4** Theories of leadership
  - 8.4.1 *The trait qualities of leadership*
  - 8.4.2 *The behavioural qualities of leadership*
  - 8.4.3 *The situational qualities of leadership*
  - 8.4.4 *Self-leadership*
- **8.5** Motivation
  - 8.5.1 *The nature of motivation*
  - 8.5.2 *Theories of motivation*
    - The needs (content) theories
      - Maslow — hierarchy of needs theory
      - Alderfer — the ERG theory (existence, relatedness and growth)
      - Hertzberg — the two-factor theory
      - McClelland — the acquired needs theory
    - The cognitive (process) theories
      - *Expectancy theory*
      - *Equity theory*
      - *Goal-setting theory*

The reinforcement theories
Types of reinforcement
8.6 Chapter review
8.7 Chapter questions

8.1 What is quality leadership?

Is leadership telling someone what to do, or is leadership asking them to do something, expecting a positive response? Leadership can be both. It can even be managing, even though managers are given responsibilities, duties and the means to carry out their tasks through legitimate power of reward and punishment and can be thought of as within the context of their legitimate position. Leaders can emerge from groups and take *charge*, not because of a given right, but because of their demonstrated ability to influence and manage people. In organizations, good leaders are seen to be individuals who influence people positively, and through utilization of resources, secure the objectives set. Thus, leadership is not an individual orientation, but a group one and therefore involves other people. Managerial leadership is what we are concerned with here.

> Leadership is the lighting of man's vision to higher sights, the raising of a man's performance to a higher standard, the building of a man's personality beyond its normal limitations. (Drucker, 1954)

> Leadership is a broad visionary activity that seeks to discern the distinctive competence and values of an organisation; to articulate and exemplify that competence and those values; to inspire, even to transform people in the organisation, to feel, believe and act accordingly. (Mukhi *et al.*, 1988)

> Leadership is interpersonal influence, exercised in a situation, and directed, through the communication process, toward the attainment of a specified goal or goals. (Tannenbaum *et al.*, 1961)

Further, Rauch and Behling (cited in Yukl, 1989) defined leadership as 'an attempt at influencing the activities of followers through the communication process and toward the attainment of some goal or goals'. This definition seems to concur that leadership is about influence. A similar definition occurs in Stoner and Freeman (1989) by Stogdill. Leadership is a reflection of a group's inner motivation towards goal or task accomplishment. The *leader* provides the appropriate environment in which group members feel most comfortable, in order to enhance performance and attain specific group goals. Without *clear* and *consistent* quality leadership, quality cannot hope to succeed (Sheehy, 1994). Quality leadership, based on an extension of the quality improvement concepts (Reiley, 1994), provides a flexible method to address the complex quality management issues facing organizations today. Quality leadership should therefore be made a strategic objective (Feigenbaum, 1993).

8.2 Strategic quality leadership — the need

Total quality management requires increased effort from everyone in the organization, in order to ensure that customers are satisfied continuously. Complementing the need for flexibility in dealing with customers, individuals also need the flexibility to operate within the defined character of the organizational mission and strategic objectives.

The balancing of an application of a directing framework, while allowing staff to make decisions, is a difficult task to achieve. This is the task of strategic quality leadership. Shared power, shared responsibility and shared commitment mean nothing if staff move in many directions and management move in another. The paradox here is that strategic quality leadership means developing everyone as leaders, not just management. Management's task has therefore changed to being a leader in learning (Long, 1993) and the quality-oriented organization is a learning organization. Essentially, management must cultivate a culture of leadership from top management through all levels in the organization (Staub, 1993), where team-based leadership measures are emphasized (Tompkins, 1993).

8.3 How leaders influence — the power stakes

Sorohan (1994) concludes that what distinguishes total quality leaders from run-of-the-mill ones, is the way they learn and share what they have learned. As a consequence, quality leaders influence people by this very nature, and that is their power — or is it? Leaders have power, but there are other sources of power, and much rests not with the leader but with the people that the leader is trying to influence. But what is power? Mintzberg (1983) and Pfeffer (1981) indicate that power is the capacity to affect the behaviour of others. In 1959, French and Raven suggested that there were six forms of power — legitimate, reward, coercive, expert, information and referent. These are further expanded in Table 8.1.

Although these six bases of power can be taken individually, there is a basic set that the other bases relate to. These are:

- Legitimate power can form the basis for reward, coercion and information power.
- Expert power can form the basis for information and referent power.

In the quality-oriented organization, legitimate power may not be used as extensively as in other more hierarchical organizations. Nevertheless, it will become apparent when problems of dysfunctional groups occur. On the other hand, expert power would actually increase in application in the quality-oriented organization. Here, the creation of motivation in subordinates, or more appropriately in peers, seems to take a greater emphasis, where coaching and

Table 8.1 Forms of power

Legitimate	Reflects the *position* an individual holds in an organization. The higher the individual in the hierarchy, the more legitimate power and therefore authority the individual has
Reward	Reflects the ability to control and manage rewards considered of value to others, e.g. financial resources, pay or promotion
Coercive	Reflects the capability of punishing others, when they have not engaged in expected behaviours, e.g. secured agreed deadlines or performed according to agreed budget arrangements
Expert	Based on the possession of expertise considered worthy in the organization. Expertise in the form of technical skills is predominant, but negotiating and communication skills are also noteworthy
Information	Information is power because anyone who controls information has that power. Managers tend to have greater information power because they have greater access to it as a *right* of organizational position
Referent	Reflects the power other people give an individual as a consequence of personal characteristics or attributes they perceive they have

stimulating rather than coercing play important roles. It seems that the quality-oriented organization would use power that is relevant to its cultural development, as a support, not a forcing element.

8.4 Theories of leadership

Leadership has become an important notion when we consider the major changes experienced by us all over the past twenty-five years or so. Now, the social conditions operating in many communities — including organizations — have changed and are changing dramatically. Customers have changed and expect organizations to do likewise. Leadership, therefore, underpins the continuing effective development of this change.

8.4.1 *The trait qualities of leadership*

It was during the First World War that leadership traits became a research focus in order for Army psychologists to choose the most appropriate individual to lead a group of people (Bass, 1981). Traits can be seen as singular characteristics that supposedly differentiate leaders from non-leaders and are considered to be qualities such as intelligence, charisma, strength, courage, self-confidence, etc. Traits were first developed as a theory at the beginning of the century and coincided with the development of Taylorism. The relationship can be clearly seen when we examine Taylorism as the development of specialized tasks, and trait leadership as the examination of specialized causes. Bass (1981) suggested that there are five correlations with leadership. These are intelligence, dominance, self-confidence, job-related knowledge and high motivation and energy. Research into traits that separate effective leaders from ineffective leaders is still inconclusive. Trait leadership theory tried to develop a means by which a leader could be selected, as it was assumed that the trait was inherent — it was something you were born with, rather than something that could be learnt.

8.4.2 *The behavioural qualities of leadership*

When trait theories failed to explain conclusively the differences between leaders and non-leaders, researchers began looking for another avenue. The development of behavioural approaches to management gave rise to the behavioural approach to leadership. Behaviours could be seen, whereas traits could only be interpreted, and at best extrapolated from behaviours. The basic assumption underpinning behavioural theory was that leadership was learnt, and therefore identifying leadership behaviour could provide a basis for teaching people how to become leaders. It was therefore suggested that leadership would never be in short supply in any given situation. To see how true this assumption is, you only have to look around you today. What do you see? Organizations clamouring for leaders — who seem to be in short supply.

Two research studies worthy of mention are the Ohio State University and the University of Michigan studies.

In the Ohio State University studies, the researchers developed a two-dimensional approach (from a list of 1000) that categorized most of leadership behaviour. The two dimensions were initiating structure (extent of structuring one's own as well as a subordinate's job to carry out set tasks) and consideration (extent of job characteristics that relate to trust and respect for subordinates' abilities and development). These can be seen in Table 8.2. Their interaction can be seen in Figure 8.1.

A manager considered being high in both dimensions, generally provided the basis for enhanced performance of subordinates and group, as well as

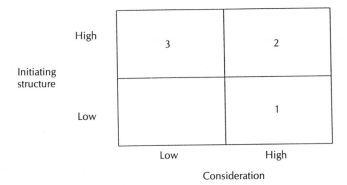

Figure 8.1 Consideration and initiating structure.

Table 8.2 Consideration and initiating structure

Consideration	Support for subordinates. Examples include fighting for subordinate rights and needs above what is expected normally, personal favours, seeing a subordinate as a peer
Initiating structure	Structuring own and subordinate job roles. Examples include controlling subordinates pace of work, ensuring that subordinates follow SOPs and being task-oriented

providing greater work satisfaction. However, this was not always the case. This has implications for quality circles and other such work groups. Not all people (subordinates) want this type of leadership and therefore will respond negatively in terms of worker performance and output.

States 1, 2 and 3 relate to the perceived effectiveness of the application of the style of leadership. A state 1 situation — high consideration and low initiating structure — results in a low perception of relationship or task-related problems. This is perhaps relevant to a professional undertaking, and reflects the autonomous nature of such posts. A state 2 situation — high consideration and high initiating structure — results in a low perception of relationship problems, but an enhanced sensitivity to task-related problems. Here, individuals sense the need for teamwork, but generally resist pressure related to task accomplishment. A state 3 situation — low consideration and high initiating structure — results in a forced mode of work practices, in which individuals are closely supervised in their work operation.

Of the different states that exist using the two combinations, state 2 is perhaps the most important in terms of a balanced output, coupled with a well-developed working environment.

In the University of Michigan studies, the managerial grid was developed using, again a two-dimensional grid. One leadership dimension was labelled employee-oriented (people-oriented) and production-oriented (task-oriented). Blake and Mouton (1964) developed this into the Managerial Grid.

The grid was created by dividing each dimension into ten equal parts — creating a grid of 100 squares. Essentially there were five important grid squares. These were:

1. 1,1 *leadership style* — An impoverished leader who exerts little influence to accomplish work-related tasks. This style cannot actually be seen as leadership at all.

    There is little concern for people or production matters. The style relates to Thomas' notion of conflict avoidance, where staying out of trouble is the primary motive.

2. 1,9 *leadership style* — Country-club style that focuses on people-oriented issues of welfare and support. The leader tries to develop a comfortable environment in which to work in, leading to enjoyable work relationships

— but not highly productive units. The conflict style here seems to be one of mediation and compromise.

3   **5,5 *leadership style*** — A centre leadership style that applied adequate task and people influences to get tasks carried out. Many leaders try to provide this sort of environment, but it is highly illusory. Production or people forces will ensure that the leader will consequently bias their activities towards production or people.

4   **9,1 *leadership style*** — Task-oriented to the almost exclusion of people. Leaders opt for this style when the external environment exacts a pressure on the organization such that survival is threatened. Consequently, internal group processes are forced to change to ensuring that each group is aligned to the needs of the organization, rather than the needs of the various groups. But, this style may reflect an inadequate balance in the leader's behavioural make-up.

For example, the individual may be conditioned by the culture of the organization to behave in this way. The focus of top management may be thus inclined to reward this type of behaviour and a people orientation may be seen as a sign of weakness.

5   **9,9 *leadership style*** — Balanced leader that facilitates integrating the task and people requirements. This is the team-oriented style of leadership, where the task and the people controlling it are as important as each other. Paradoxically, this style is being used to great advantage in organizations that are facing survival pressures — much like 9,1 leadership style. However, nurturing and developing people will mean that although the external pressures may increase, the people in the organization are capable of meeting the challenge. This is the challenge for the quality-oriented organization. The style is characterized by attempting to modify both the organizational and group goals and to align them to a single direction. By doing so in a constructive way, both the organization and the group benefit.

What seemed to have been missing here though, was the reaction to the environment in which both leader and 'led' are situated. However, cultural factors have rarely been considered worthy of attention. It is to these situational elements that we now turn.

8.4.3 *The situational qualities of leadership*

Situational orientation suggests that leadership styles are more appropriate in one situation than another. But what are those conditions and their relevant styles? Barrow (1977) indicated that the variables of the task performed including complexity and technology used were significant, as was the style of the leader's immediate supervisor, the developed span of control, the given importance of the total task and the strength of the organizational culture. A number of approaches have been developed for evaluating the key variables for situational leadership. These approaches include the following.

Autocratic–democratic Continuum Tannenbaum and Schmidt (1973). At the one extreme, the manager makes decisions for the subordinate, whilst at the other extreme, the subordinate shares in the decision-making role.

Whilst this is not exactly a leadership model, it is indicative of the leadership style that provides for varying decision-making capability, and it is the situation (the operating conditions) that predominates in the change from one part of the continuum to another.

The following indicates the leadership continuum style.

1 — Leader makes decision and announces it
2 — Leader *sells* decision
3 — Leader presents ideas and invites questions
4 — Leader presents tentative decision — subject to change
5 — Leader presents problem, gets suggestions on solution and makes decision
6 — Leader defines limits and asks group to make decision
7 — Leader permits subordinates to function within limits as defined by the leader

This model has limitations. Tannenbaum and Schmidt reduced the meaning of autocratic away from its expected meaning. That is, it did not necessarily mean that the leader applied punitive measures or was totally task-oriented. However, there is some similarity between this model and the Vroom and Yago's model of managerial decision-making practices. The model therefore attempts to provide some understanding of the breadth of managerial activity with respect to the relationship between the leader and the subordinate. Leaders are expected to evaluate their needs, the subordinates needs and the needs of the situation, when determining which leader behaviour to adopt.

The quality-oriented organization would encourage going beyond the right-hand limits set by the model. In this way, quality management practices provide a basis for mutual negotiation and localized decision making, without the need to involve the manager or leader. This is the basis for the operation of quality circles and other such group quality practices.

Situational Leadership Theory Hersey and Blanchard (1978). This theory focuses on the followers, rather than the leaders. The theory was first postulated in 1969 as the 'Life-Cycle Theory of Leadership' (Hersey-Blanchard, 1969), but was later changed to the situational theory we now know. Their theory rests squarely on the assertion that leadership must be contingent upon the maturity of followers (the degree of responsibility followers take for their own actions through psychological — motivational elements — and job-related knowledge and skills).

Their model addresses four specific leadership styles:

1. *Telling* — Control rests with the leader and the leader dictates what, when, where and how people carry out their job-related tasks (high task orientation — low people interaction).
2. *Selling* — The provision of both a directive-type behaviour, as well as supporting the individual in carrying out their job-related tasks (high task orientation — high people interaction).
3. *Participating* — Sharing of decision-making and job-related information with the leader facilitating and coaching rather than directing (low task orientation — high people interaction).
4. *Delegating* — Subordinate requires no directing and little leader support (low task orientation — low people interaction).

This approach is akin to the grid developed by Blake and Mouton as follows:

Telling — 9,1 leadership style
Selling — 9,9 leadership style
Participating — 1,9 leadership style
Delegating — 1,1 leadership style

Path–Goal Theory (developed by House, 1971). This considers that an individual's leader behaviour is acceptable to subordinates when they perceive an immediate or direct source of satisfaction. This satisfaction, according to House and Dessler (1974) comes from two areas — first, that satisfaction is contingent of effective performance and second, that effective performance comes from the coaching and supporting style of the leader. It would seem that initiating structure is appreciated by subordinates when the task itself is unstructured, whilst consideration is deemed helpful when the subordinates task is ambiguous.

The following leader behaviours are identified by the Path–Goal theory:

1. *Directive* — Providing direction to subordinates as to what is required, when and to whom, to what standards of output and the requisite reward as a consequence. Generally this is seen to be best applied to immature employees, in the sense of experience in regard to the task at hand.
2. *Supportive* — Providing concern for the needs and welfare of subordinates through positive relationship building. The leader considers the psychological basis for the employees' needs and correspondingly develops a positive attitude in the employees of their importance in the task completion.
3. *Participative* — Providing an environment in which subordinates are seen as partners in the management of a group by encouraging and consulting them before making decisions.
4. *Achievement-oriented* — Providing challenging outcomes from both goals and the process used to achieve those goals. The leader indicates and develops increasing confidence in the abilities of the employee to accomplish set tasks.

The Fiedler Contingency Model (Fiedler, 1967). This suggests that group performance is dependent upon how much the situation gives control to the leader and the leader-style offering. An instrument he developed, now called the least-preferred worker scale (LPC), measures whether an individual is task- or person-oriented. The scale is illustrated below, as eighteen bipolar adjectives:

Pleasant _____ Unpleasant

Problems of scale interpretation and the meaning associated with the labels are well documented. However, lately an interpretation suggests that the LPC measures a sort of motivational hierarchy, where the focus is on the task, rather than the relationship between workers. Fiedler assumes that the preferred leadership style needed by an individual is fixed and cannot be changed, and thus is not situationally dependent.

Consequently, determining their preferred style means determining how they want to be treated by leaders. According to Fiedler, there are two ways to improve leader effectiveness — first, fit the leader to the situation and second, change the situation to fit the leader. Structurally, the second option is actually easier to attain than the first. However, both are difficult to assess and apply using Fiedler's model.

Situational leadership is not supported by many theorists, but its promise for developing as a model still has to be tested effectively. Other situational leadership theories include:

Cognitive Resource Theory (Fiedler, 1986). This is really an extension of Fiedler's work and considers the cognitive resources — intellectual abilities and task-related and other demonstrated technical competences. This takes aspects of leadership theory back to trait theory application. Other researchers support this viewpoint (Bass, 1981).

Transactional and Transformational Leadership. The distinction between transactional (motivation of subordinates to expected levels of performance) and transformational (motivation of subordinates beyond what is expected) leadership was first made by Burns (1978). Transactional leadership is closely related to the Path−Goal theory developed earlier, as in this theory transactions are seen as the key to performance.

Transformational leadership on the other hand is about inspiring subordinates to excel in their performance, to take risks, to innovate and to achieve beyond expectancy. This form of leadership is closely related to the expected operation of a quality-oriented organization. Transformational leadership can only occur effectively after an efficient base is created through transactional leadership. This form of leadership is predominantly carried out at a distance from the actual working conditions or operation of transactional leadership — as charisma is seen as a major influence on its effective implementation, although, not everyone seems to think so (Howell and Frost, 1989).

Vroom and Yago (Yetton) Model (1988). This has been developed and used as a basis for a normative (indicating what should be done, rather than what is done) leadership style. The Vroom and Yetton model of 1973 was revised into the Vroom and Yago model in 1974, and further again in 1988. In 1988 it was developed into a computerized application that managers could use on-line. The model attempts to guide managers to use the correct managerial leadership process effectively in the perceived circumstances facing the leader and the group.

The model uses three different management styles that reflect the situation requirements — group, individual and task. Vroom and Yetton have taken these three basic styles and broken them further into five styles:

- **A**   Autocratic: has two variations — AI and AII
- **C**   Consultative: has two variations — CI and CII
- **G**   Group: has one variation — GII

Table 8.3 illustrates the definitions attached to each style and variation.

Managers can become proactive in the way that they determine how to lead their groups. In order to make this rather prescriptive model easier to use in the real world, Vroom and Yetton produced a series of twelve questions that seek to evaluate an effective decision path for any problem facing a leader. These twelve questions are illustrated in Table 8.4.

The model also tries to ensure that managers consider more than just the task or more than just the relationships in the group. The computerized version provides positive feedback in the form of graphics about the advice that the model suggests.

8.4.4 *Self-leadership*

These theories are fine as academic inquiries, but what about the practical applications? What have these theories to do with the everyday operation of a plant, an office or a school?

The application of quality management practices reduces hierarchical levels in an organization. Consequently, this forces the levels of responsibility increasingly downwards. In the quality-oriented organization this has entailed individuals and groups taking on greater leadership roles both for themselves and for the people they work with. In some instances it has resulted in the total application of leadership in groups, rather than by individuals leading a group. Thus, quality leadership demands a new kind of *tough leadership* (Parka, 1993). Self-leadership means actions and activities that are designed to provide individuals with the necessary skills, education, commitment, knowledge and motivation to carry out task-related tasks on their own, and to know when they need to consult and co-ordinate group effort in order to accomplish group goals.

Managers take great risk when implementing self-leadership practices in organizations, as these very activities provide the backdrop to the managerial

Table 8.3 Five styles of Vroom and Yetton

Style	Explanation
AI	Leader makes the decisions for the group, using information available to the leader at that time
AII	Leader generates information from subordinates — may or may not indicate the purpose of the request for information. The information is used to solve a problem determined by the leader and subordinates do not take part in any evaluation of alternative strategies that could be implemented. Leader makes decision
CI	Leader determines problem definition and shares the ideas with subordinates individually. Leader makes decision that may or may not reflect subordinate input
CII	Leader determines problem definition and shares the ideas with subordinates in a group. Leader makes decision that may or may not reflect subordinate input
GII	Leader determines problem definition and shares the ideas with subordinates in a group. The group develops alternative solutions to the problem and attempts are made to elicit a consensus of agreement. The leader role here is one of facilitation, rather than as a manager. The group makes the decision, in which both the leader and the group are committed

Table 8.4 Twelve questions of Vroom and Yetton

How important is the technical quality of this decision?
How important is subordinate commitment to the decision?
Do you have sufficient information to make a high-quality decision?
Is the problem well structured?
If you were to make the decision yourself, is it reasonably certain that your subordinates would be committed to the decision?
Do subordinates share the organizational goals to be attained in solving this problem?
Is conflict among subordinates likely over preferred solutions?
Do subordinates have sufficient information to make a high-quality decision?
Does a critically severe time constraint limit your ability to involve subordinates?
Are the costs involved in bringing together geographically dispersed subordinates prohibitive?
How important is it to you to minimize the time it takes to make the decision?
How important is it to you to maximize the opportunities for subordinate development?

changes that will ultimately occur. Some individuals want leadership for themselves in order to lead others — this is the traditional approach. In the quality-oriented organization, individuals want leadership primarily for themselves and then — if needed — for the group. Consequently, the adoption of a different attitude to leadership, commensurate with quality practices, ensures the survival of the group and the individuals in that group.

In the quality-oriented organization, it means that leadership is becoming a requirement, not from others, but for ourselves. Self-leadership is therefore taking on greater importance. Self-leadership is a requirement of quality circles, quality planning groups, etc., and is becoming a personal need. Self-determination is fine, but self-leadership means the development of confidence, skills — personal and technical, developing a sense of control over one's workspace and understanding the effect of personal contribution that can be made. It also means controlled individual, and therefore group, transformation.

Deming (1993) suggests that transformation is profound knowledge of the system, variation, and the psychology of change, which means that education and training, taking risks and self-control are major pillars of transformation and therefore self-leadership. The quality-oriented organization ensures that people are able to self-lead. It means ensuring that they are trained and educated, resourced and, above all, encouraged to identify with this leadership style. In some way, quality management theory has provided a very positive orientation to the general working life. Self-leadership means knowing when and how to participate when decision making needs to be done, and become committed to the outcome.

Self-leadership has been labelled in a similar vein as self-control and self-management. These are misnomers. Self-leadership goes far beyond these other limiting notions of individual management, as it is a philosophy (Manz and Sims, 1990) and management practice that leads to all-round increases in individual, group and organizational performances in order to satisfy customers' needs and wants. In this way self-leadership becomes an ongoing internal process and leads to real personal empowerment. Implications of self-leadership include:

1. Reduced need for managers and their specific job requirements.
2. Heightened need for continuous training and education programmes.
3. Reculturization of management practices to take advantage of the new leadership style.
4. Increased risk of failure if workforce does not engage in the necessary commitment.
5. The increased timeframe needed to develop such a self-leading culture.
6. Increased co-ordination that is necessary at all levels in the organization.
7. Reconciling the requirements of quality management systems like BS EN ISO 9000.
8. Dealing effectively with the need to share appropriate information.

8.5 Motivation

8.5.1 *The nature of motivation*

What is motivation? How are individuals motivated? What can managers do to enhance the motivational environment? These are aspects that will be treated in this part of the leadership chapter. Leadership cannot be effective unless a manager understands the basis for motivating individuals. This is a basic requirement in the application of quality management practices, because it is individuals who actually carry out the work of an organization and managers have adopted these practices in order to increase individual performances. Consequently, motivation theory has a very important role to play in the development of leaders or managers of quality-oriented organizations.

Motivation can be seen as the force generated by individuals towards accomplishing goals — internal and external — which in itself creates behavioural inertia. Motivation, therefore, gives direction to an individual's orientation that generates a tendency to persist (Bartol and Martin, 1991). Motivation is an internal drive that reflects the influences of internal and/or external stimuli.

8.5.2 *Theories of motivation*

Individual performance is likely to be a function of the task, complexity of task process, individual ability, internal motivation and external environmental factors. Management has to attempt to manage all factors that are considered external to the individual — as these can be controlled to some degree. Early approaches attempted to evaluate the motivational energy of an individual as an isolated entity. For example, many theorists think that Taylor underpinned his scientific management theory on the basis that an individual could be motivated solely by money, which when applied to the work environment resulted in increased output.

However, as we have already seen, this was not quite correct and the resultant output may be attributed to increases in other factors, not solely borne by the motivation for money. Incentive schemes were formerly developed by management and their only real control of the situation was through the application of wage increases. This was no doubt acceptable to the general workforce, as the economic conditions at the time were such that money however earned was necessary for survival. However, many workers feared increased production led to increased lay-offs and a reduced workforce. Other factors seemed to impinge on the straightforward application of money as a motivator. Consequently, the focus given by management gave rise to a bias in management theorists such that they may have been misled.

Since the worker did not seem to be totally motivated by economic gain, what does motivate the worker? Social theories of the 1930s also gave rise to the application of motivation theories that reflected those developments, in particular the experiences of Elton Mayo at the Hawthorne plant.

It was suggested that giving people attention, attempting to increase worker decision-making powers, especially in relation to direct worker responsibilities, and evaluating the effect of group process and group incentive schemes (Steers and Porter, 1987), could provide a more integrative basis for motivating individuals. These attempts were directed to the group rather than the individual and as a consequence failed to deliver the promise that by increasing the socializing influence on individuals, their motivation also increases.

The needs (content) theories

Needs can be seen as the basic inner drive of an individual. It is the basic notion applied to all marketing effort that subscribes to the application of the marketing concept. In this instance, needs drive wants, and wants create demand. Consequently, need theories have tried to determine what can be managed in order to influence an individual's drive positively, although mostly for organizational purposes. In this respect, needs theories have also become known as the *content* theories.

Maslow — hierarchy of needs theory

Maslow (1954) developed what has become known as the *hierarchy of needs*. This theory was the first of its type actually targeting the internal reason for the individual's drive. Although, as we will see, this does not mean exclusion of the external environment. Maslow identified that there are five needs. These needs include:

1. *Physiological needs* — such as basic food, water and shelter. In the work environment, these translate into a basic living wage or salary, and the workspace in which to operate.
2. *Safety needs* — such as home or community security and safety. In the work environment, these translate into a safe working environment, job security and other negotiated benefits.
3. *Social needs* — such as belonging, acceptance and affiliation to family and other groups and individuals. In the work environment, these translate into relationships with peers, supervisors, managers, etc., formal and informal work groups.
4. *Esteem needs* — such as a good self-image and the desire to have contributions recognized and accepted by others. In the work environment, these translate into recognition of good performance by peers and other co-workers, especially superiors, and the development of status through recognition — name tag on an individual office — that affects the viewpoint of these individuals.
5. *Self-actualization* — such as higher internal ideals and reaching full potential. In the work environment, these translate into the need to challenge and develop creativity, with the consequent growth in potential and capability when successful.

These needs are illustrated in Figure 8.2.

Maslow suggests that each need, from the bottom up, must be satisfied to some degree before we can take the next step upwards. However, the orientation of an individual to concentrate solely on a given level is a rather simplistic view. Many individuals seem to be required to work on all these levels at the same time. Also, these identified needs are related in some way and seem to be dependent, rather than independent as was claimed. Maslow's theory has been tested many times and its inadequacies have highlighted the fact that some individuals will focus on one need satisfaction at the cost of the others, i.e. some monks may

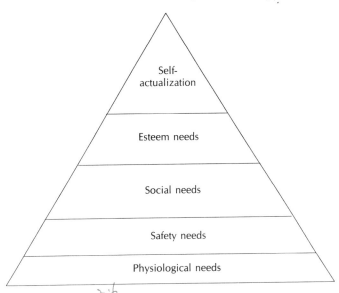

Figure 8.2 Maslow's hierarchy of needs.

focus on high ideals, yet lead reasonably deprived lives physically. Some research has indicated that some individuals concentrate on satisfying two or three of the five needs (Wahba and Bridwell, 1976), a form of clustering need satisfaction. The influence of the needs may also change in response to the human lifecycle, where different needs are focused on at different periods in an individual's life.

Alderfer — the ERG theory (existence, relatedness and growth)

In 1972, Alderfer suggested an alternative to Maslow's theory and its determined shortcomings. The ERG theory differed from Maslow's, with only three categories of needs, instead of five.

Existence compares to the Maslow's physical and safety needs; relatedness compares directly to Maslow's social needs; and growth compares to Maslow's higher-level needs of esteem and self-actualization. One of the major differences between Maslow's hierarchy and Alderfer's is that Alderfer's seems to be able to be more easily verified and therefore provides a greater relationship with reality.

It must be said that, as motivational theories, both Maslow's and Alderfer's applications are rather limited and thus retain their relative prescriptiveness. Their actual effect in a quality-oriented organization is still in conjecture. Knowledge of individuals and their actual needs is very important, but exactly how these two theories provide a basis for measuring and evaluating changing individual needs, and their effect on individual and group performance are points for discussion.

As an extension, Alderfer suggested the application of a *frustration–regression* principle. This suggests that when we are confronted continuously by failure of satisfaction of a higher level, then the individual's response is to regress and concentrate on the next-lower level. In this way, lower-level satisfaction is reinforced. In this regard, higher-level needs are then ignored. This is rather simplistic, an addendum to the main part of Alderfer's theory. However, possible mechanisms that provide for a forced short time in the lower level, and then attempt to further satisfy a higher-level attainment are likely.

Hertzberg — the two-factor theory

The outcome of Hertzberg's two-factor theory is somewhat related to Maslow's needs theory. Hertzberg distinguished higher- and lower-level components of what seemed to him to motivate an individual. The basis for Hertzberg's research was conducted in 1958, where he asked 200 engineers and accountants to describe what situations satisfied and motivated and what did not satisfy them and left them unmotivated. Consequently, Hertzberg derived his two-factor theory from the responses. Satisfiers were determined as factors such as achievement, responsibility, work itself, recognition and growth.

Dissatisfiers were determined as factors such as pay (salary), working conditions, supervisors and fringe benefits. These are illustrated in Figure 8.3.

Hertzberg indicated that no matter what the content of the hygiene factors was, they could not of themselves provide motivation, but did provide a means to prevent dissatisfaction. Only attention to the factors recognized to create motivation will ensure that an individual is motivated to perform effectively.

Subsequent researchers testing Herzberg's ideas have not been able to provide the theoretical link between motivating situations and their consequent motivating factors. Also, some researchers (Steers and Porter, 1987) have indicated that dissatisfiers can become motivators and that hygiene factors can of themselves lead to satisfaction.

This means that different people respond in different ways and that the target audience should be considered. The theory may be useful and very wide in application, but needs to be modified to accommodate this. However, what seems significant is that unions — possibly without exception — focus on the hygiene factors and do not seem to be concerned or interested in the motivational factors, as these are thought to be the domain of management.

McClelland — the acquired needs theory

This theory is firmly based on the notion that our inherent needs are acquired as we live our lives. That is, it is based on learning through life experiences. Therefore, it is the reaction to the external environment stimuli that derives the acquired need. For McClelland there are essentially three needs — achievement, affiliation and power. These are tested using the Thematic Apperception Test

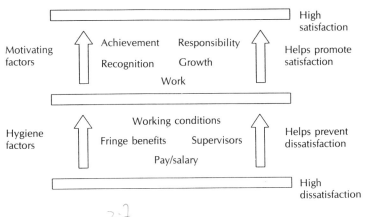

Figure 8.3 Hertzberg's two-factor theory.

(TAT), which is essentially a test using pictures. Respondents develop stories while viewing a given ambiguous picture or pictures, and the consequent evaluation of the written theme provides an evaluation against the scores attributed to each of the three identified needs.

Managers need to be aware of the bias given one of these three needs, as it provides some basis for their use and effectiveness in an organization or group. For example, a manager with high achievement needs does not necessarily focus on developing the same sense in their subordinates. The individual is likely to concentrate on their own tasks and achievements. They are seen as moderate risk takers, as the challenge is required to give them the sense of achievement. However, extreme challenges are generally not undertaken because of the high risk of failure (McClelland, 1985). High achievers are considered to be useful in situations that require creativity and innovativeness. High-affiliation individuals tend to communicate more effectively with others, and consequently favour tasks that require the development and maintenance of relationships. This may be in a training or teaching dimension.

High-power-oriented individuals have a developed desire to influence and control others, and situations. In this regard McClelland differentiates between the need for individual and organizational power. The best managers are individuals who have a high need for organizational power and therefore put the needs of the organization first. These individuals are more likely to want to organize other people's work and therefore control resources and processes.

Steers and Porter (1987) suggested that a successful manager has a moderate to high need for organizational power, a moderate need for achievement and at least a minimum need for affiliation. This contradicts much of the basic management principles, in that affiliation is almost a constant requirement in organizations — at any level — and that individual power would seem to provide middle to top management with the ability to control the parts of their organization that they have responsibility for.

The cognitive (process) theories

These motivation theories differ from the needs theories by attempting to evaluate the internal mental thought processes that influence behaviour. These theories are essentially driven by the assumption that an individual weighs-up the possible outcomes of expended effort in achieving a particular goal, and determines *a priori* whether to go ahead or not. Major theories that will be considered here are the expectancy, equity and goal-setting theories.

Expectancy theory

Vroom suggested that we consider three issues prior to expending effort to perform a given task; these form the base for the expectancy theory. These are:

1. *Effort–performance expectancy* — This suggests that we evaluate the probability that the efforts we expend will lead to the acceptable performance of the given task. This involves an evaluation of factors such as own ability, knowledge, resources and the context and critical nature of the given task.
2. *Performance–outcome expectancy* — This suggests that we evaluate that successful performance of the task will lead to expected outcomes. This involves evaluating remuneration (extrinsic rewards), achievement and status (intrinsic rewards).
3. *Valence* — This is the value we place on the anticipated outcomes, both positive and negative. If the anticipated rewards interest us or are seen as worth the effort, then the valence will be high. On the other hand, if the anticipated rewards do not interest us, then the valence will be low. In this respect, an individual will reflect not only on the organizational issues, but their personal life also.

Where one of the issues is considered low, then the likely motivational state of the individual would be low. Consequently, motivation is likely to be high only when all three elements are considered to provide a high response. This may mean increased negotiation — organizational or personal — in order to get the perceived motivational values increased.

Researchers have found difficulty in making direct comparisons between individuals in the same situation — mostly because individuals do not seem to consider the same aspects. Consequently, this means that a full analysis of the expectancy and what it can offer managers is not altogether clear. Pinder (1984) indicated that the theory is not actually applied to the individual, but to providing a means to predict an indiviual's response in a given situation. This would seem to narrow the application of the theory in practice.

Equity theory

Equity means just that — equity. We view situations and outcomes relative to what we and others get for the same expended effort. Equity means evaluating these situations and comparing what others get with what you get. If what you get is deemed as of equal value to what others get, then there is a balance, which most people strive for and are most happy with. Unfortunately, not every situation provides the basis for a balanced outcome for everyone. There are winners and losers.

Problems of the application of this theory include the differing perceptions related to the situation and the value or utility placed on the compared outcomes. Motivation in this theory is channelled to alleviating any discernable inequities found. Since many inequities can be found by simply shifting the comparison other, then the inequity energy creates a motivational tension. As the perceived equity increases, so does this motivational tension.

What seems to create great problems in the application of this theory is the almost limitless inequity situations that can arise — the change of comparison other, the different perceptions held by supervisors and other managers — all result in an overwhelming motivational tension that needs to be reduced. Reduction in perceived equity would likely result in first maximizing outcomes, then attempting to change factors in the situation, then cognitively altering or changing one's own value system and then possibly leaving the situation altogether if none of the previous elements are feasible. For example, if an individual has developed high professional standards and is denied the ability to practice those standards because of suggestions of cost from a supervisor (whilst others can), then the individual may leave the job because the motivational tension becomes so great, especially if the individual cannot negotiate out of the perceived inequity. Increased communication in groups and organizations may be needed in order to reduce the possible effects of equity theory, so that individuals know the rules they are working under and have full knowledge of the basis for decisions.

Goal-setting theory

Setting goals is inherently a means to create motivation because they generate an awareness of what must be accomplished, by when, with what and by whom. Consequently, an individual's involvement is known and the performance requirements are set. Coupled with this is the requirement for monitoring and evaluation of performance, which when implemented provides reinforcement for performances met.

Locke *et al.* (1981) indicate that goal-setting is a strong motivational tool. In the quality-oriented organization, much use would be made of this tool, as it is the basis for developing and evaluating quality plans. Consequently, its use in organizations can have considerable effect, both for organizational, and individual performances.

The reinforcement theories

Reinforcement theory takes no consideration of the thought processes that determine behaviour. Consequently, it determines behaviour outcomes in isolation and considers that the behaviour reinforcement of positive behaviour attributes and characteristics is the focus. Reinforcement theory can be viewed as manipulative and dogmatic.

Much of reinforcement theory is based on the evaluation of the environment (the cause) and the individual's reaction to it (the effect). In this respect, positive consequences are seen as more likely to be repeated than negative consequences (seen from the individual's viewpoint). Positive and negative consequences are personal. For example, some consequences are seen as positive to one individual and negative to another. Reinforcement theory therefore can be used to modify present behaviour into a different future behaviour. The trick is to find what will motivate an individual to make those changes.

Types of reinforcement

There are four types of reinforcement. These are:

1. *Positive reinforcement* — A reinforcement technique that focuses on increasing the repetitiveness of a given behaviour, by positively rewarding the demonstrated act of that behaviour. The positive consequences usually involve pleasant acts such as praise, recognition, money. But it could be construed as positive when punishment is given. The reaction is totally personal.

2. *Negative reinforcement* — A reinforcement technique that focuses on increasing the repetitiveness of a given behaviour, by making the individual engage in a given behaviour in order to stop another, unwanted stimuli. The individual is given the choice of either accepting a change to the new, desired behaviour or is further subjected to the unwanted behaviour. Many managers engage in negative reinforcement, as this seems to be the predominant reinforcing style in organizations, especially in small ones.

3. *Punishment* — A reinforcement technique that focuses on decreasing the repetitiveness of a given behaviour, by providing negative consequences. The ability of a manager to legally sanction staff puts that behaviour firmly in the category of punishment, but any change of behaviour could be construed as punishment in cultural terms. Punishment brings with it the reinforcement of the notion of power and control.

    It may also bring negative feelings against the punisher, which in itself may endanger good relationships if not used as had been previously conditioned or in respect of agreed equity limits.

4. *Extinction* — A reinforcement technique that focuses on decreasing the

repetitiveness of a given behaviour, by reducing or stopping the positive consequences of that behaviour. Extinction leaves the control with the individual perpetrating the unwanted behaviour, and as such is therefore not proactive. Consequently, a change in behaviour may not occur, as the actual reinforcer is the attention given to the individual by attempting to ignore the demonstrated behaviour.

Reinforcement, as a theory, has been applied rather more successfully than many of the previous theories. In terms of real behavioural changes, positive reinforcement — whilst being proactive — seems to command respect. However, reducing a human being to the status of a machine or animal in terms of contemplated reaction, does not engender the notion of organizations containing thinking, innovative, flexible and resourceful people.

8.6 Chapter review

Leadership is an attempt to influence individuals positively to carry out tasks and responsibilities that the leader deems important. Leaders do this by using power effectively. Power generally exhibits itself in one of six forms — legitimate (reflects the position of an individual in an organization), reward (reflects the ability to control and manage rewards), coercive (reflects the capability of appropriate sanctions), expert (based on the possession of knowledge or expertise that the organization needs), information (reflects the ability to control and manage information considered necessary for an organization's operation) and referent (reflects the power other people give to an individual).

There seem to be three categories of theory of leadership. These are:

1. *Trait* — Singular characteristics that differentiate leaders from non-leaders. Examples include intelligence, charisma, strength and self-confidence.
2. *Behavioural* — A visible, learnt process that underpins the basis that leaders can be taught. Studies have indicated that two dimensions — initiating and consideration — are important criteria for evaluation of leadership style. These criteria are taken without consideration of the external environment. Examples of the applicable theories include the Ohio State and Michigan Universities and the managerial grid.
3. *Situational* — Suggesting that situational factors are important because leadership styles are more applicable in one situation than another. Various theories have been applied to the leader,

the followers and the environmental context in which both have to operate. Examples include the autocratic–democratic continuum, situational leadership theory, the path–goal theories, cognitive resource theory.

Self-leadership is developing as a leadership theory, and as such has implications for the management of self-led groups in quality-oriented organizations.

Motivation can be seen as the force generated by individuals towards accomplishing goals — internal and external — which, in itself creates behavioural inertia. The nature of motivation suggests that an individual's performance is likely to be a function of the task, complexity of task process, individual ability, internal motivation and external environmental factors. Theories of motivation include the need, cognitive and reinforcement theories. Need theories explain that we are continually attempting to satisfy different needs, including Maslow's physiological, safety, social, esteem and self-actualization, Alderfer's existence, relatedness and growth, Hertzberg's two-factor theory and McClelland's acquired needs theory of achievement, affiliation and power. Cognitive theories explain that the thinking process determines the way in which we behave. These theories include Vroom's expectancy theory; equity theory; and goal-setting theory. Reinforcement theory postulates that behaviour results from reactions to the external environment. It can be divided into four areas — positive reinforcement, negative reinforcement, punishment and extinction.

8.7 Chapter questions

1. Define the term leadership.
2. Apply the six bases of power to a group that you belong to, and evaluate your own *power* position.
3. Explain the development of leadership theory.
4. Outline the significance and use of Tannenbaum and Schmidt's autocratic–democratic continuum.
5. Discuss and apply the Vroom and Yetton model of decision making to a group. Discuss the impact that computer technology may have on its subsequent use.
6. Discuss the various motivation theories. Highlight your motivational experiences when applying the theories to a group that you belong to.

CHAPTER 9

Group dynamics

CHAPTER OBJECTIVES

Define a group
Describe the characteristics of a group
Explain the types of groups found in organizations
Evaluate group development and its implications
Evaluate group effectiveness and group efficiency
Explain the requirements and need for team building
Evaluate the content and effects of conflict in an organization
Discuss the need, configuration and requirements of communication in groups
Explain the communication process

CHAPTER OUTLINE

- **9.1** Introduction
- **9.2** What is a group?
- **9.3** Characteristics of a group
- **9.4** Types of groups found in organizations
  - 9.4.1 *Formal groups*
  - 9.4.2 *Informal groups*
- **9.5** Group development
- **9.6** Group effectiveness and efficiency
  - 9.6.1 *Group inputs*
    - Group composition
    - Group appeal
    - Group roles
    - Group size
  - 9.6.2 *Group process*
    - Group norms and conformity
- **9.7** Team building
- **9.8** Conflict
  - 9.8.1 *Managing conflict*

9.9   Communication
    9.9.1 *Why communication is important in groups*
    9.9.2 *Types of communication*
        Verbal communication
        Non-verbal communication
    9.9.3 *The communication process*
    9.9.4 *Group communication configurations*
        Centralized configurations
        Decentralized configurations
9.10 Chapter review
9.11 Chapter questions

9.1 Introduction

The development of groups or teams in organizations, sports activities and any other human involvement where individuals are required to work together, is seen as a necessary factor in today's society. So let us explore what some theorists have stated about groups.

9.2 What is a group?

A group is defined as having in its membership two or more people working towards common goals. Berne (1963) describes groups as 'aggregations which contain at least two classes of people, one — the leadership, and two — the members'. Groups are seen as extremely important in organizations. French and Bell (1984) make the assumption that 'in today's organisations much of the work is accomplished directly or indirectly through groups'.

In most organizations, the structure will be such that many groups carry out the necessary functions of its operation. It is because of this that management are turning to groups to ensure the delivery of products that continually meet customer specifications, within ever-increasing cost and competitive constraints.

9.3 Characteristics of a group

1   Common purpose or goals to achieve.
2   Shared sense of identity — both internally and externally.
3   Participation — interaction opportunities.
4   Cohesiveness.

5   Some semblance of structure — leader and led.
6   Sanction capability in relation to its membership.

These characteristics can also be attributed to organizations themselves.

9.4 Types of groups found in organizations

Groups can essentially be divided into formal and informal groups.

9.4.1 *Formal groups*

A formal group is an officially sanctioned group that has been formed for a specific purpose by the organization. These are generally divided into functional and special purpose groups.

- *Functional groups* — Functional groups are formal groups set up in order to carry out the everyday tasks of the organization. They consist of managers and subordinates who work together to accomplish the goals of the organization; each department will have functional groups and the whole department could be considered a functional group. It is a matter of degree.

  The manager of one group may be a member of another group, and the manager of that group may be a member of another group higher in the hierarchy. In this way, Likert's (1961) linking-pin concept can be seen to be applied, providing one of the bases for co-ordination through the organization. In this way, organizations can be seen as structures with formal groups connected by linking pins. This is illustrated in Figure 9.1.

- *Special purpose groups (SPGs)* — A special purpose group is a formal (permanent or temporary) group created to support the normal organizational

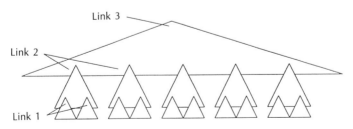

Link 1 — Link from worker to supervisor
Link 2 — Link from supervisor to middle manager
Link 3 — Link from middle manager to top manager

Figure 9.1 The linking-pin concept.

operation. The permanent or temporary nature can refer to the output, not to formation characteristics. An example is the SPG formed to manage a new building project, or the organizational move to new premises. The membership may or may not be the same (depending on the technical nature of the project), but the focus will be different each time. A further example is perhaps quality improvement groups, set up to solve quality-related problems in a defined area of an organization. A standing committee is another example. Temporary SPGs include project teams or task forces.

9.4.2 *Informal groups*

An informal group is one formed by employees for their own purposes and interests, not the organization's. Informal group membership will include members of functional groups or SPGs. The nature of the informal group means that the boundaries and conditions affecting the management of functional groups do not apply. For example, the chairperson of an informal group may be a shop-floor worker, even if its membership contains the plant manager. Informal groups are seemingly underestimated in their ability to influence organizational affairs and their usefulness needs to be recognized. There are essentially two types of informal groups — interest and friendship.

1. **Interest** — An employee interest group is created to provide a focus of action for interests of concern. Concern could be in the form of health — sports club creation; or in the form of a hobby — railway enthusiasts; or they could be in the form of internal or external political groups who are pursuing interests that affect them in their work.
2. **Friendship** — This is a grouping to *formalize* a means to meet employee social needs. These include the use of social gatherings, social trips to major

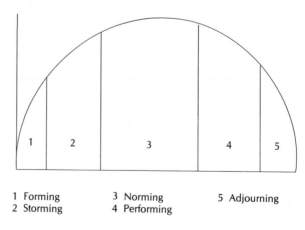

1 Forming 3 Norming 5 Adjourning
2 Storming 4 Performing

Figure 9.2 Stages in group development.

cities, holidays and other special interests, such as golfing and theatre activities.

Both these types of groups provide a relaxing, enjoyable way for employees to socialize.

Trends are such that these groups are becoming important connections in the organizational network. It would seem that more information and faster transfer through the organizational grapevine (see communication later in this chapter) can sometimes be given through the formal organizational groupings.

9.5 Group development

As change today becomes the norm in organizations, so is the acceptance of changing groups or group compositions. This has implications for group development. Groups seem to move through various stages of development in their operational pursuits, whether formal or informal. Tuckman (1965) believes that groups move through five stages of development. These stages are illustrated in Figure 9.2.

The stages are:

1 **Forming** — Members determine the ground rules for their alliance in carrying out their tasks and responsibilities. Members work out how to interact and test their influence and power affecting the group dynamics. Where the new group consists of individuals who have never worked together before, the hierarchical situation will influence the structural development of the group.

   In this case, members who occupy higher positions will influence the group operation more than members who occupy lower positions in the organization. Kanter (1983) called this the *seductiveness of the hierarchy*. Task orientation may be slow at first because of the consequent focus on group process and group relationships at this time. Standards of operation are tabled for discussion but probably not dealt with, or solved effectively.

2 **Storming** — This stage is where individuals investigate whether they can influence the group activities and test different outcomes in process as well as in task orientation. New alliances will be attempted as changing situations begin to point to the development of power individuals or subgroups. The central theme seems to be the continual conflict associated with leadership of the group. This is the stage when group fractionalization may cause the group to become permanently dysfunctional and may actually cause the group to break up.

3 **Norming** — Once the leadership and control issues have been *amicably*

resolved (though they may come back later), group norms are worked on and developed. This could mean the acceptance of a formal or informal leader. There will become a consensus about how the group will operate. There will still be conflict, but this is resolved on an ongoing basis and possibly resolved with time. Group identity becomes more important than individual squabbles.

Operational matters have been fine-tuned so that members have a clearer understanding of what is expected of them. Paradoxically, this in itself reduces the personal tension that existed in the previous stages.

4   **Performing** — This is the expected outcome of the group formation, to be able to operate as effectively as possible, although the group may not actually perform as the organization intended. Therefore, the group could be effective, but in another direction. This stage is characterized by the focus of attention on the tasks set for the group. It is likely that problem-solving skills, developed in the previous group development stages, can now be employed on group tasks to great effect. Since members have now worked together for a reasonable length of time, the learning curves have shortened and personal relationships have been developed to a degree where trust and support are enhanced.

5   **Adjourning** — This is where the members prepare for group dissolution and seek other alliances with other individuals or groups. This is easier when the task duration for the group is finite, as in a building project. Where the group task is ongoing, this stage creates similar problems to those that existed during the storming stage.

It seems all groups move through these various stages, although some may not actually experience the changes as they do. Some groups only get to the storming stage, others move through a complete cycle. The main concept here is that group development is essentially time-based. It requires time to develop norms of task activity, as well as norms involving personal relationships. It would also seem that the focus of developing personal relationships is by far the most influential aspect of early group dynamics, and not the task.

Formal groups therefore develop their internal characteristics and productive capabilities over a period of time and it seems reasonably accepted that groups move through various stages of development. It is important for a group to understand its stage of development, because performance is heavily influenced by it. Bass (1965) developed a model that provides a basis of understanding task group behaviour and orientation. This model is discussed below:

Stage type 1: Orientation.

Behaviour characteristics
1   Beginning of communication patterns.
2   Development and knowledge of interdependencies among members.

3 Acquaintance with the structure and goals of the group.
4 Expression of expectations.
5 Mutual acceptance among members of each of the others as a group.

When the members of the group formed, the initial interactions generally involved preliminary discussions of the group's objectives, becoming acquainted with each other's knowledge and abilities and developing a plan for future interactions and activities.

Stage type 2: Internal problem solving

Behaviour characteristics
1 Problems arising from stage 1 are confronted and attempts at solutions made.
2 Increased interpersonal conflict because individuals bring to the group unresolved problems relating to different feelings toward authority, power and leadership, etc.

Unless problems or conflicts are confronted and solved to the satisfaction of the group members, the performance of the group will be adversely affected and the group may never advance beyond this stage.

Stage type 3: Growth and productivity

Behaviour characteristics
1 Group activity is directed almost totally to the accomplishment of the group goals.
2 Interpersonal relations within the group are marked by increasing cohesion, sharing of ideas, providing and getting feedback, exploring actions and sharing ideas related to the task to be done.
3 Individuals feeling good about being part of the group, leading to emerging openness and satisfactory performance toward goal accomplishment.

Stage type 4: Evaluation and control

Behaviour characteristic
1 Focused evaluation of individual and group performance. This is accomplished through the adherence to group norms, strengthening interdependencies and structure and various feedback mechanisms.

Each stage in the development of a group needs to be accomplished before the group can move on to the next. At each stage boundary 'pinches' and 'crunches' exist that must be solved for the group to effect positive change.

Models like those discussed above indicate that there are differences in orientation and performance of groups which essentially depend on the length of time they have been working together.

9.6 Group effectiveness and efficiency

What constitutes group effectiveness? One particularly important aspect is performance. Group effectiveness is generally measured using this as the sole criterion — an output condition. Since internal customers will have both inputs and outputs to and from other internal individuals or groups, then performance constitutes a major issue, especially for efficiency. Characteristics that are likely to affect group performance are generally placed in two areas — inputs and group process.

9.6.1 *Group inputs*

These include group composition, appeal, roles and the intended size of the group.

Group composition

Hackman (1987) indicates that managers need to consider various issues when choosing group members. Issue number 1 is that members should have technical expertise to carry out relevant group tasks. If not, the political machinations, learning curves and outside allegiances may create process problems. Some management decisions are made without due consideration of this, and group entropy results (a state of disorder). Since much of the initial group development is through interpersonal development, then issue number 2 is that members are trained in interpersonal skills. Issue number 3 indicates that heterogeneity, where all members think similarly (resulting in less conflict), is not acceptable for high-performing groups. A diversity of views and skills is not just required but is considered a must.

Group appeal

Many group members belong or attempt to belong to groups for more reasons than just to carry out work-related activities. High-performing groups are gateways to career advancement. The group performance is therefore seen as a stepping stone that induces esteem, recognition and achievement. Consequently, some group operations have greater appeal than others. High-profile projects, that are short-term oriented and with a considerable degree of success, will undoubtedly lead to a greater competitive influx of group membership. Promotion provides great impetus to individuals and how they manage the groups they belong to or want to belong to. Effective management of this aspect will ensure a group membership that is committed.

Group roles

The managerial roles, as discussed by Mintzberg (1980); interpersonal, informational and decision play an important part in the development of a group. These roles may not be assigned to any individual, but the behaviours attached to their use may be quite evident, e.g. leadership may be given to an individual only because of certain special skills, abilities or information access. The informational roles of monitoring, disseminating and/or spokesperson can be carried out by all members, but some may be more dominant and effective in some roles than others. Decision-making roles are usually practised by everyone, although some members may lack the confidence to implement group decisions without persuasion. Where a dominance is seen in a particular role, other group members may tend to allow that individual to administer that role almost exclusively, e.g. disturbance handlers smooth the group process, reduce conflict and clarify different points of view. These various roles were expanded in Chapter 1. Other roles include task, maintenance and personal roles; Adair (1983) developed a model that integrated the task, maintenance and personal roles and needs within a group, referred to as group functions that have to be performed. Essentially, the roles need to be carefully assessed so that each group member is completely satisfied for their part with the three functional roles. Where an individual lacks task solution ability, then they need to develop maintenance roles to accommodate this problem, since other group members' support is required.

Group size

Shaw (1981) investigated how the size of groups affected performance. The number of individuals in a group compounded interaction and communication channels. It seems that small groups — between five and seven — produce the most effective decisions, whilst groups between eleven and fifteen produce the most committed decisions. However, the larger the group, the more difficult it is to gain consensus. Decisions are made on majority rule or on the balance of power, and equity reduces and failure rates increase. It is quite surprising that, when facing dissolution, groups generally become unanimous, especially when no political bias in outcome accrues to any group member or members. Effectiveness depends on the size of the group compared to its stage of development. Size affects how quickly the group can move through the various stages towards the performance stage. The larger the group, the slower the movement and therefore the greater the likelihood of a dysfunctional group occurrence. The larger the group initially, the more difficult communication patterns that are set up and translated into task and problem solution, rather than interpersonal relationship solution.

9.6.2 *Group process*

Some groups at the same stage in their development perform better than other groups, even of the same size. Why is this? One of the main reasons seems to be a good group process. Individuals may feel more comfortable, are socially acceptable, are equitably challenged and feel safe. The group feels good to belong to. But good process brings about a negative side.

If people feel good about themselves in a group, then there is generally great resistance to change and the way they are used to working just in case there are changes to the experience and feelings attached to the group operation. Janis (1972) indicated these problems of inward-looking groups.

Factors that affect group process and its consequent effectiveness include group norms and conformity.

Group norms and conformity

All groups, both formal and informal, have sets of standards for the behaviour, attitudes and even the perception of their members. These standards are called norms, which are shared expectations for what is appropriate and inappropriate behaviour, and what members should and should not do.

All groups not only exhibit norms, they also have mechanisms by which conformity to these norms is accomplished, i.e. ways in which group pressures are brought to bear on members who stray too far from those developed group's norms. The content of the norms varies from group to group, as does the degree to which the norms are explicit or implicit. Groups also vary in the form which normative pressures take, in the intensity of these pressures and in the targets of these pressures. An indication of the type and content of the norms that a group can develop is:

1   Equal productivity and quality of output.
2   Individual members are expected to pull their weight.
3   Rotating leadership.
4   Punctuality at any meeting.
5   Rotating location.
6   Fostering an environment whereby group members listen to others in the group.

The following model, contained in Figure 9.3, illustrates situations for four cases that affect group efficiency and effectiveness.

- CASE 1 — LOW EFFICIENCY/LOW EFFECTIVENESS

The situations when case 1 could arise are:

1   The start-up of a group.
2   A major shake-up of the group, say, resource reshuffle, or introduction of a new process, rules, structure, technology, etc.
3   New leader.
4   Threat of change — uninformed or insufficient official information leading to group speculation.
5   Perceived unfair treatment of group or individuals in group.

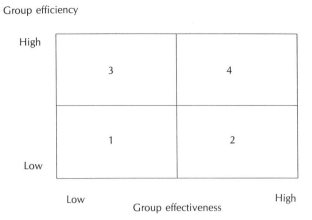

Figure 9.3 Group efficiency and group effectiveness.

6 Group is pressurized by other groups or leaders.
7 No or little group direction.
8 Unwillingness of the group to work together.
9 Politics of the group insufficiently characterized and multi-directional.
10 Characterized by the move from case 4 to case 1.

- CASE 2 — LOW EFFICIENCY/HIGH EFFECTIVENESS

The situations when case 2 could arise are (the group needs to become effective before efficiency rises):

1 Group has been working together for a small period of time.
2 Group goals are only just taking precedence over individual goals.
3 Positive feedback is being maintained within the group.
4 Individuals are highly motivated.
5 Group process of rules, programmes and communications, etc., is developing.

- CASE 3 — HIGH EFFICIENCY/LOW EFFECTIVENESS

The situations when case 3 could arise are:

1 The task definition and support is well developed and may interfere with group business (external requirements).
2 The group has direction and therefore has developed inertia.
3 Group goals have total dominance over individual goals.
4 Group conflict is dealt with according to the developed procedures, rules customs, etc.

- CASE 4 — HIGH EFFICIENCY/HIGH EFFECTIVENESS

The situations when case 4 could arise are:

1 Group is highly task-orientated.

2   Individual goals are suppressed for the good of the group.
3   The group feels the task, personal gain and the group status are enhanced.
4   The group perceives itself as being very cohesive.
5   Communication between group members is enhanced.
6   The group takes on a more collaborative style for dealing with group conflict.
7   Group politics of status, power and authority are becoming suppressed, even for the leader. This suppression will evolve at a later stage in this case to cause fractionalization problems with group members.

These cases relate directly to the stages of group development.

9.7 Team building

Since group development has a profound impact on group performance, it makes good sense to move groups as quickly as possible to the performing stage. In order for this to occur team building has been suggested. An acceptable modern definition is given by French and Bell (1984) 'activities the goals of which are the improvement and increased effectiveness of various teams within the organisation'. Team building is also defined by Szilagyi and Wallace (1980) as 'a planned event with a group of people who have or may have common organisational relationships and/or goals that is designed to somehow improve the way in which they get work done'.

Woodcock and Francis (1979) determines that team building is the 'process of planned and deliberate encouragement of effective working practices while diminishing difficulties or blockages that interfere with the team's competence and resourcefulness'. Team building, then, is a technique which allows processes and relationships to be developed to effect positive change and enhanced performance from the group by securing individual participation. It does this by developing a sense of group identity, rather than an individual disposition. Team building can focus on permanent or temporary work groups.

The type, effect and length of the team building intervention differs with different organizations, the team structures, philosophies and culture, the people concerned and the objectives and type of task to be performed, e.g. different interventions are needed to build teamwork within base worker groups than professional worker groups, due to the different expectancies, values and capabilities of each group. Whatever the type of intervention, there is a need to focus on all members of the group and just a few. This corresponds to the system-wide intervention of organization development programmes.

Not all situations require team building. The situations where this is appropriate are (Buchanan and Huczynski, 1985):

1   When co-operative working is likely to produce a better end result (in terms of speed, efficiency or quality) than working separately.
2   When the amalgamation of work into joint tasks or areas of responsibility would appear meaningful to those involved.

3   Where the joint task requires a mixture of different skills or specialisms.
4   Where the system requires fairly frequent adjustments in activities and in the co-ordination of activities.
5   Where competition among individuals leads to less effectiveness rather than more.
6   Where stress levels on individuals are too high for effective activity.

The above furnish basic criteria in which to apply to differing situations. If a negative response is received from any of the above, then team building is not going to be an effective measure to increase performance in individual members of that group.

Teams that are seen as effective exhibit:

1   Cohesive attention to common goals.
2   Each team member understanding those goals.
3   Individual performance measurement.
4   Mutual expectations from each group member regarding performance and behaviour.

In addition McGregor (cited in French and Bell, 1984) states that effective teams exhibit the following:

1   The atmosphere tends to be relaxed, comfortable and informal
2   People express both their feelings and ideas
3   Conflict and disagreement are present but are centred around ideas and methods, not personalities and people
4   The group is self conscious about its own operation
5   When actions are decided upon, clear assignments are made and accepted by the members

To determine whether a given situation would benefit from a team building strategy would require the application of one or more of the diagnostic methods available from many literature sources. Unfortunately, as determined by Woodcock and Francis (1979) there is a distinct lack of suitable models to assist managers and human resource specialists in deciding whether team building is likely to be a sound investment. Nevertheless, there are other methods available that will allow the effective development of teams. Quality improvement tools are a means by which a team can gain not only from the results it achieves, but also from the increased team member control of their working environment through participation.

9.8 Conflict

The first casualty of war is — Innocence
The second casualty of war is — Respect
The third casualty of war is — Trust

Conflict is the process which begins when one party perceives that the other has frustrated, or is about to frustrate, some concern (Thomas, 1976). Further, Miner (1978) suggests that conflict has its origins in differences in objectives, interests, efforts, approach, timing, attitudes, and so forth. Much of it is consciously recognized by the participants and intentionally produced. Evan (1965) indicates that conflict is inevitable. Sometimes it is even desirable, because it can reflect a vital organization composed of knowledgeable people who care about creative performance.

So conflict would seem normal to group process and therefore must be recognized and managed, not brushed aside and ignored. The best groups manage conflict because, even though the group works together, they are still individuals who think independently but operate together. The generation of conflict — and it will be generated where resources are scarce and individuals are competing for those resources — can have both a positive and a negative effect on an individual's performance, motivation, competence and enthusiasm in task completion. Understanding the mechanics of managing conflict in groups will allow individuals to assess their own conflict position, that of their group or organizational peers and determine strategies for ensuring conflict which does not denigrate the individual's interaction in group or organizational affairs.

Conflict is also seen as necessary — but not too much. Too little and individuals may be hiding the generated conflict and group performance may not be developing to its optimum. Too much conflict, and too much energy is directed at taking sides and using power management, rather than working on the group tasks. In between there is an optimum. Figure 9.4 illustrates this.

Conflict has a number of causes (Walton and Dutton, 1969). These include:

1   *Task interdependence* — Since processes are interdependent, the individuals who work in them are therefore interdependent. Morgan's (1986) machine

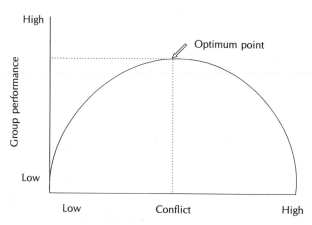

Figure 9.4 Conflict and group performance.

metaphor of input—process—output provides some basis for understanding this. Consequently, failures in one part of a process affect the desired effectiveness of another. One type of task interdependence is sequential interdependence in processes. This results in direct conflict concerning failures in process output. Another type is seen as mutual interdependence, where process and output are dependent on another process controlled by another group. This makes inter-team building a necessary management deliberation.

2   *Different key objectives* — Groups that have been together a long time develop different ways of operating and may in fact develop group-think (Janis, 1972). Their orientation may become out of synch with the department or the organization as a whole and create conflict, not only by virtue of their different operation, but also by the different directions they are pursuing.

3   *Rewards structures* — Structures that give better rewards to one individual for the same quality and quantity of output create notions and feelings of discontent through injustice and unfairness. Developing equitable pay systems can be a daunting task, human resource managers need to ensure that whatever system they develop does not create conflict in this way. This is especially true when dealing with the development of reward structures that cross different groups and departments that are interdependent.

9.8.1 *Managing conflict*

There are many ways of managing conflict in a group. Three major ways of managing conflict are changing the situational factors, developing superordinate goals and interpersonal.

1   *Situational factors* — These factors include managing resources in such a way as to reduce conflict, such as increasing resources — financial, physical and human. Changing the mix of resources to reduce conflict may actually cause more conflict in the long run, as individuals become reinforced by their conflict actions.

2   *Developing superordinate goals* — Reducing internal group conflict by focusing support on greater ideals and objectives than the actual group requirements. It is a process where all energies are raised to solve problems created outside the group, so that internal problems and differences are laid aside. This is a temporary measure, as the external orientation will give way very quickly, to the ever-present intra- and intergroup conflict.

3   *Interpersonal* — Managing interpersonal conflict is managing conflict, since it is only individuals who create conflict in the first place. Thomas (1976) introduced the notion that managing interpersonal conflict involves the use

of one or more conflict styles. These styles of conflict management are:

- *Avoidance* — where individuals seek not to make a decision about a situation or problem in order to suppress conflict, even if they were capable of making that decision, backed by suitable levels and type of resources.
- *Competition* — the imposition of a preferred solution by a party who has the power to do so. A weakness of this technique is that one who dominates may well get no co-operation from those he or she dominates.
- *Accommodation* — acceptance of other parties, position without gaining any new position or ground.
- *Compromising* — the identification of a solution that to some extent meets the needs of both parties.
- *Collaboration* — involves finding a solution that completely serves both parties. It requires clever thinking, thinking outside the bounds of two mutually exclusive alternatives, trust, the exchange of accurate information and a belief that the process can produce two winners, not just one.

9.9 Communication

Communication can be seen as 'the transfer of information from the sender to the receiver with the information being understood by both sender and receiver' (Koontz *et al.*, 1986). Also, communication is 'one person giving a message to another. Communication is essentially a dialogue: It takes two or more to communicate' (Adair, 1973). Irvine (1970) indicates that communication is an exchange of 'information, interpretations, ideas, opinions, and decisions between human beings'.

Communication is everything. In marketing, management, social situations, our lives. If we do not communicate, we are isolated. But what is communication? Communication is defined simply as the exchange of meanings through messages. Essentially, it means the exchange of messages between individuals or groups of individuals with the fundamental purpose of achieving a common understanding. Communication is required in order to influence others. Managers particularly need to communicate in order to direct and control effectively.

9.9.1 *Why communication is important in groups*

In the quality-oriented organization, communication is an absolute necessity. Furthermore, much of the basis for quality management rests squarely on the principle of shared information. Information cannot be effectively shared if available information is not communicated in a timely and accurate fashion. Consequently, much effort must be given to the task of enhancing the communication process. Initially this could take the form of managing information requirements in groups and through a process of assimilation this could be extended to other parts of the organization. Interdependency makes information sharing necessary. Interdependency ensures that managers communicate.

The development of effective communication processes cannot be achieved overnight. Cultural determinants have implications, but mostly to resist rather than assist change. In this respect, communication developments must be optimized to the present and possible near future states of the organization. An example here is the shop-floor worker given the responsibility for attending board meetings as a gesture from top management towards increased worker participation. The reality of the situation is that the worker cannot communicate with the board members very well, mostly because their language is different from the workers. This means that discussions will not include the wholehearted approach of the shop-floor worker and consequently the initiative will be short lived.

Given training, and a longer time to assimilate, the shop-floor worker may be able to address the board as an equal, but it is not surprising that there are many failures.

Communication is not just telling someone; communication is about explaining. This is part of the positive philosophy of effective training programmes. Explaining what needs to be, doing it and then explaining what went right and what went wrong is much more acceptable than just telling someone they have failed or that they failed to match acceptable standards. The latter is an example of one-way communication — quality control. More positive arrangements suggest coaching and explaining. In this respect, this requires the development of two-way communication skills — a proactive and improvement-oriented approach.

In the quality-oriented organizations, groups seem to prefer decentralized networks, rather than centralized ones — irrespective of the task. The basis for this seems to be that the group is focused on developing an effective process, and as such — because of an improvement and problem-solving philosophy — consider the task as being secondary. This does not imply that the task is seen as unimportant, but it does indicate that good process in that environment will always ensure that skills are developed to bring an effective solution to the task at hand. The communication pattern of these types of groups is worth investigating. Predominantly, there is horizontal communication. This is where lateral exchanges occur between intra- and intergroups — either within the department or in other parts of the organization. What is perhaps significant is that quality-oriented organizations engender this type of communication throughout.

The organization effectively becomes a matrix structure, as the quality practices force changes to the overall structure, such that it reduces the levels of management. Greater communication therefore derives from restructuring as a consequence of an increasing quality orientation.

9.9.2 *Types of communication*

Managerial communication, like any social form of communication, has two major types — verbal and non-verbal. Each is an effective medium, but predominantly, managers use verbal communication more, although the effectiveness of such communication may rest solely on the non-verbal attributes.

Verbal communication

This is the written or oral use of words to communicate meaning. Written communication includes both internal — memos, reports, data and operational elements of performance — and external — market research and intelligence reports, customer requirements, stakeholder reports and marketing brochures and promotion efforts. Much internal verbal communication involves short-term communication such as meetings and memos. Other media include the use of telephone or other technology communication devices. In this respect, much of this type of communication is on a one-to-one basis, where the exchange is interactive — whether in meetings, in the corridor or on the telephone, etc.

Written communication tends to be relatively long in its development, impersonal and slow in providing feedback. However, complex issues that need deliberation are better confined to the written medium, as much information is lost in the oral medium. Oral communication tends to be fast, with feedback almost instantaneous within the confines of the communication content and can be used to confirm, more often than to discern. Major disadvantages include being time-consuming, and difficulties arise on when to terminate such interactions and the risk of focusing directly on the issues as control of focus of the discussion becomes difficult.

Non-verbal communication

This is where communication occurs through anything other than by oral or written means. Effectively, it is communication through behaviours and symbols that determine the meaning given to them in the communication by the individual. Expressions, body gesture and posture derive meanings that individuals instantly recognize and take notice of. These are the conditioned responses that may have developed prior to an individual learning to talk. There meanings become interpreted on the basis of reinforcement. Consequently, non-verbal communication is cultural communication and as such, attributes different meanings to similar behaviours and/or symbols. Oral communication cannot be achieved effectively without some form of non-verbal communication. Much non-verbal communication is used to provide a confirmation key or credence to the oral message. In this way, managers can enhance the communication process if they know the audience and their communication biases in respect of non-verbal communication.

Non-verbal communication includes kinesic behaviour (body gestures, etc.), proxemics (the amount of space an individual needs to see between them and another individual in order to feel comfortable), paralanguage (voice quality and tone that indicates how something is said, rather than what is said) and object language (objects providing statements about the individual, e.g. materialism).

Management preferences depend on their managerial level in the organization. Mintzberg (1983) suggested through a study of five top managers that 78 per cent of their time is spent on oral communication. How does this compare with first-line supervisors? It depends on whether the organization is quality-oriented or not. If it is quality-oriented, there is evidence that there is an increased use of written communication, rather than oral communication. This is borne out when we consider that the generation of data and its traceability to a given action requires this. Consequently, lower management is increasingly preoccupied with ensuring adherence to stated written procedures.

9.9.3 *The communication process*

The communication process can be seen in Figure 9.5.

The various components are encoding, sender, medium that the messages flow through, the receiver and decoder. This is one-way communication. In an interactive situation (two-way communication), the above process is repeated and becomes cyclic in nature.

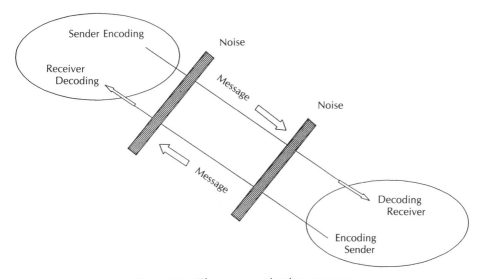

Figure 9.5 The communication process.

The *sender* is the initiator of the message and must first encode the message prior to sending. This encoding translates the intended message into a meaning that is relevant for the receiver. To do this, the individual translates the messages into symbols that can be understood by the other party.

The *message* is any translated symbol that is being conveyed to the receiver. In this regard, the medium needs to be considered in order to ensure the most appropriate method for the audience.

The *receiver* is the individual that takes the intended message and decodes it into a meaning that they can understand. It is here that meanings can differ between the intended message sent and the actual one decoded. This places great emphasis on ensuring that the symbols that will be used to decode the message are understood by the sender and that the message and the medium used consider these. What should be achieved is that the shared meaning of the sender and receiver is the same — otherwise communication is lost and time and effort wasted.

Noise is anything that interferes with the transmission of the message. Noise may cause the generation of other symbols in the message and therefore when decoded will give a different end-result to the communication.

Feedback is an initiated response to the decoded message. This may be in the form of an oral, written or just non-verbal message. The proximity and access to the initiator will determine what can be done in the form of feedback. Feedback provides the basis for developing and continuing to keep shared meaning. If there is dialogue, and more importantly continuous dialogue, there is likely to be enhanced communication.

One-way communication means just that, the message is transferred one-way only. This does not allow for confirmation or for further inquiry. Two-way communication, on the other hand, means that the interaction will generally result in a richer information exchange and will therefore lead to fewer ambiguous meanings being attached to messages.

9.9.4 *Group communication configurations*

Most work in an organization is carried out by groups of individuals. Various communication configurations have been developed over the years and each has its advantages and disadvantages. These generalized configurations can be broken down into centralized and decentralized configurations. These can be seen in Figure 9.6.

Centralized configurations

These include the reverse V, Y, wheel and chain networks. The individual occupying the top of the reverse V controls the communication channels from both forks. Likewise, in the Y network, the individual at the top of the stem controls the communications to each stem. In the wheel the individual at the

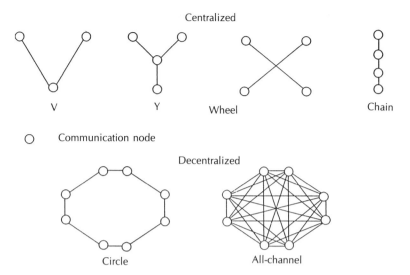

Figure 9.6 Communication configurations.

centre controls the communication channel to each of the other members of the group. In the chain network, communications between members of the group must flow through one individual who then controls the flow of information.

Decentralized configurations

Decentralized networks offer greater flexibility in operation and therefore provide a more effective environment in which to communicate. Consequently, decentralized configurations are more widely adopted in the quality-oriented organization. Teamwork and problem solving all require the nurturing and reinforcement of greater communication. The circle therefore provides a middle step to full decentralized working practices contained in the all-channel network configuration. Unfortunately, all-channel configurations are by their very nature limited to small groups and predominantly those in close operational proximity.

However, the tasks that groups in these configurations have to perform must be taken into account. Where the task is repetitive and relatively simple, with defined procedures that can be followed effectively, then a centralized network will seem to be more useful.

9.10 Chapter review

A group is defined as having in its membership two or more people working towards a common purpose or goals, and having two classes of individuals — the leaders and the followers. Groups bear characteristics that include a shared identity, cohesiveness, structure and sanction giving. Groups are very important in organizations, as they carry out much of its activities.

The types of groups found in organizations are formal — officially sanctioned groups formed for a specific purpose by the organization, and informal — groups formed by employees for their purposes and interests.

Groups are seen to change over time, because of changing organizational requirements as well as changing characteristics relating to the group dynamics. This group development can be viewed in five stages — forming, storming, norming, performing and adjourning.

Group effectiveness is generally measured in terms of output. Characteristics that are likely to affect group performance are generally placed in two areas — inputs (group composition, appeal, roles and the intended size of the group) and group process (group norm development and conformity). There are four stages that link group effectiveness and group efficiency and these are associated with the stage of group development.

Team building is planned and systematic activities designed to enhance group processes and to provide a fast-track development to take the group to the performing stage. Team building therefore allows processes and relationships to be developed in order to effect positive change and subsequently enhanced group performance. However, too great a success in team building may actually harm external relationships. Therefore, a balance must be struck between building a group to perform effectively and the development of independency that participation requires.

Conflict is the process when one party perceives that the other has frustrated or is about to frustrate some concern. Conflict is the norm in organizations, and as such, needs to be managed in order to get the best out of. Conflict can arise as a consequence of task interdependence, different key objectives and differing reward structures. Three ways of managing conflict in a group or organization are to effectively manage the situational factors (resources — physical, financial and human), the development of superordinate goals and internal interpersonal factors.

Communication is required in groups in order for data, knowledge and opinions to be expressed fully. An effective communication process ensures that information is shared and made available so that individuals can make effective decisions. Managerial communication, like any social form of communication, has two major types — verbal and non-verbal. The management of these forms is necessary for management and for the workers, as decisions made by workers can affect the decisions management can make.

9.11 Chapter questions

1. Define a group.
2. Describe the characteristics of a group.
3. Explain the purpose and types of groups found in organizations.
4. Evaluate group development and its implications for managing groups in an organization.
5. Evaluate the implications of group effectiveness and group efficiency.
6. Explain the requirements and need for team building in a group, in a department and in an organization.
7. Evaluate the content and effects of conflict in an organization.
8. Explain why communication in groups is important.
9. Discuss the impact of the various types of communication patterns on group performance.

CHAPTER 10

Human resource management

CHAPTER OBJECTIVES

Define the term human resource management (HRM)
Evaluate the relationship between HRM and TQM
Describe the process of HRM planning
Explain recruitment and selection processes
Evaluate training, education and development issues
Explain performance appraisal
Evaluate compensation issues
Describe the impacts of workforce relationships and HRM
Discuss the impacts of quality circles on HRM practices

CHAPTER OUTLINE

- 10.1 Introduction
- 10.2 Definition of human resource management (HRM)
- 10.3 HRM and TQM
- 10.4 HRM planning
- 10.5 Recruitment
- 10.6 Selection
- 10.7 Training, education and development of employees
- 10.8 Performance appraisal
- 10.9 Compensation
- 10.10 Workforce relationships and HRM
- 10.11 Quality circles
- 10.12 Chapter review
- 10.13 Chapter questions

10.1 Introduction

Human resources is possibly the most inadequately managed area in an organization. This may be attributed to human capital being so inconsistent and unpredictable, unlike machines, equipment or possibly finance. Measuring the quality of machines is relatively simple, but how do you measure the quality of staff? The provision of an environment that positively promotes people in terms of skills and competencies is therefore difficult. Essentially, human capital requirements have moved in a complete circle. Prior to the Industrial Revolution, craftsmen ruled the employment stakes. Today, similar patterns of skills, competencies and job control are required. The application of on-the-job quality control practices has focused the quality-oriented organization on exploring and using these developments. Consequently, developing quality practices means developing people — and this is where an effective human resource management (HRM) system becomes imperative.

10.2 Definition of human resource management (HRM)

HRM is the process of designing workforce measures and activities in order to enhance the efficiency and effectiveness of organizational performance. In this respect, quality and HRM are seen to be in tune. Both are concerned with ensuring that organizational goals are met in the most effective way. HRM, as far as human resource (HR) managers are concerned, is about the application of functions and tasks that relate to recruiting, selecting, training and educating the workforce — including managers, other roles of union negotiation and the application of personnel practices that reflect legal and moral issues are also generally included. HRM is a very important aspect of management in a quality-oriented organization. HRM managers therefore have greater power than in traditional organizations — whatever their size — because of the impact they can make on the human resource.

10.3 HRM and TQM

Issues of managing a workforce increase when we consider upgrading the culture of an organization to take on board quality management philosophies and practices. The major key to effective quality practices is the management of human resources. O'Dell (1986) indicated the differences between the traditional and the quality-oriented approach to managing human resources. These can be seen in Table 10.1.

Table 10.1 Comparison of quality orientation

Criteria	Quality-oriented HRM	Traditional HRM
Phiosophy	Teamwork and shared understanding and commitment	Individually oriented — reward for individual work
Quality objectives	TQM orientation in every area and level of organizational activity	Production-control-oriented
Employee involvement	High — people-oriented culture	Low — system-oriented culture
Education and training	Multi-skilling orientation	Development of skills for a specific job
Reward structure	Formally owned and administered jointly by managers and workers	Management owned and administered
Structural orientation	Decentralized	Centralized

This dichotomy is not as clear as the diagram indicates. However, the stereotyping of the different cultures is quite marked, and is therefore acceptable. It is perhaps more useful to think in terms of a continuum in which the presented dichotomy lies at each end. Many organizations therefore, will exhibit some established organizational behaviour between these two extremes.

The quality-oriented HRM process can be seen in Figure 10.1.

This process is necessarily internally oriented, but externally directed. Since HRM is about people, then determining the effect of structural changes in the form of new responsibilities, tasks and skills requirements means understanding the likely impact of present and new staff requirements (if any) and all training and educational requirements that follow as an outcome of this. This causes changes to the compensation package offered to staff, in order to maintain effective workforce relationships. The process is deemed cyclic, as the need to develop continuous improvements in the HRM process is seen as vital when considering one of the best resources an organization has — its staff.

TQM can be viewed as human-relations-oriented, whereas traditional HRM seems to be systems-oriented. Human relations includes aspects of group and organizational processes, leadership and motivation, etc. Systems includes job design, engineering, production facilities and control of performance, labour relations, etc. The quality revolution, therefore, has had a major effect on how management approach human relations within the workplace. Gone are the days when demarcation lines prompted great conflict between management and workers.

Today, the quality-oriented organization is specifically focused internally on marrying effectively systems and human relations, and externally, at ensuring communication to and from suppliers and customers.

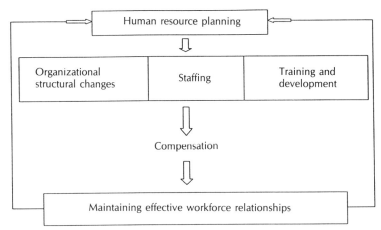

Figure 10.1 The quality-oriented human resouce management process.

10.4 HRM planning

Planning has become a major issue in the management of human resources. It affects how the organization responds to changing customer, industry, legal and competitive requirements. General planning and quality planning theory were considered earlier in this text. Those techniques have an application here too. Planning what type and quantity of human resources will likely be needed in the next twelve months is a little more difficult than forecasting material or equipment needs. However, there are techniques that have been developed in order to make this easier.

What then needs to be planned? Figure 10.2 illustrates the various components of HRM planning.

The *external situational analysis* would include an evaluation of past and present economic, legal, social and political trends and factors that could affect the future management of human resources.

Further evaluations of demographics, such as, population trends, could also occur. Specific SWOT analysis could reveal changes to situational factors that bear on the occupied and potential markets and competition therein, and also the possible effect of new technologies.

The *internal situational analysis* would include present and future evaluations of processes, production line and staff requirements. An evaluation of the effectiveness of policies and practices that affect quality-related activities would be a minimum and an absolute necessity. Appraisals of staff and their skills would also need to be implemented. This is seen as a rare event in many of today's organizations — whether quality-oriented or not. Individuals possess skills that are just not used and therefore much waste occurs. The internal situational analysis

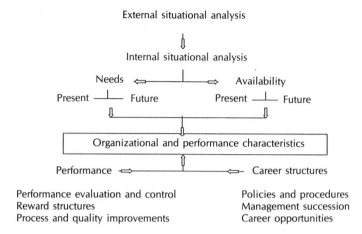

Figure 10.2 HRM planning process.

would indicate the level of gap between what was required by the workforce and what the workforce actually contained. Much has to be given to the greater use of staff, than at present time.

As an effective outcome to the above two analyses — internal and external — a forecast of human resource requirements is developed. Generally a gap exists between what a company needs and what it has. It is here that recruitment and selection procedures have become an important element in the management of human resources.

A two-tier system seems to provide the basis for much of the internal information needed for the internal evaluations. These include organizational — career management, policies and procedures, individual opportunities and performance characteristics — performance evaluation, reward structure and process improvements.

We will now explore the techniques that can be used to provide a more effective workforce arrangement in an organization. The first of these is recruitment.

10.5 Recruitment

Recruitment is the process of developing a set of organization-oriented activities that seeks to promote to and engage the most suitable candidate with the abilities, competencies and attitude required to carry out a given job-related task. Successful HR planning, HR forecasts of supply and demand, coupled with an effective job design capability, should bring about a successful recruitment campaign. Forming the basis for a recruitment plan is the development of specific forms of staff numbers, skills and competence mixes, knowledge requirements and the expected job experience level. This means that HR planners must have an intimate knowledge of job design techniques, the operational requirements of the organization — goals and strategies — and be able to communicate all this with agencies inside and outside the organization. At a time when costs of recruitment are escalating, it is prudent, if not very smart, to engage with the workforce you have at present because of one major element — demonstrated loyalty to the organization.

Among additional characteristics tagged onto candidates is that in a quality-oriented organization, teamwork, problem-solving skills and the ability to develop or apply the concepts of statistics are all necessary, whatever the job description and the intended level of recruitment. Recruitment can come from many sources — employee contacts, newspaper and journal advertisements, or just plain unqualified canvassing.

10.6 Selection

A second aspect to consider is selection. *Selection* is the process by which one or more candidates are chosen who best meet the stated job-related conditions on offer. The steps in the selection process are illustrated in Figure 10.3.

The first part in the process is to screen prospective candidates. This is done in terms of the written reply received by the organization, directly from the candidate. This first part is extremely important to prospective candidates as it affords them only one chance to progress to the next stage. In the United Kingdom at present, upwards of 80 per cent of prospective applicants are screened out at this stage. What is not altogether understood is that some candidates are screened out on ineffective criteria; for example, a screener's hunch. Next,

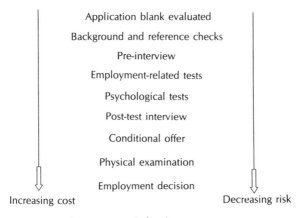

Figure 10.3 Selection process.

Table 10.2 Selection techniques evaluation

Predictor	Validity*	Accuracy ranking	Frequency of use ranking
Cognitive ability tests	0.53	1	10
Job tryout	0.44	2	11
Biographical inventory	0.37	3	4
Reference check	0.26	4	3
Experience	0.18	5	2
Interview	0.14	6	1
Training and experience Ratings	0.13	7	8
Academic achievement	0.11	8	7
Education	0.10	9.5	5
Occupational interest	0.10	9.5	6
Age	−0.10	11	9

* This is the representation of the validity coefficient, where a correlation between predictor and job performance can take values between +1 and −1. Correlations of +1 and −1 provide a perfect correlation, whereas 0.0 signifies no correlation

Printed with permission from Dakin, How to identify and assess human potential: The place of psychological testing, *New Approaches to HR Development Conference*, Wellington, NZ, May 1989, Institute for International Research, Auckland, New Zealand

references are generally checked in order to ensure that candidates are suitable for the position on offer.

A preliminary interview is carried out in order to provide suitable qualified candidates with the necessary information about the process of selection they will experience.

Employment and psychological testing is carried out next. This provides a basis for job (specific job context activities) and attitude (instruments vary between organizations) related factors to be evaluated. A warning needs to be given here. Tests have been notoriously ineffective in the way that they predict job success. Job-specific tests are much better correlated than attitudinal tests. Dakin (1989) indicated the following breakdown in terms of the validity of eleven predictors contained in Table 10.2.

Table 10.2 indicates that interviewing (0.14) — which is the most popular means for selection — is actually much more inferior than cognitive tests (0.53). So why are interviews so expressly used? Research has indicated that the validity of interviews can be increased through appropriate structuring. However, it does suggest that reliance on one aspect is insufficient. Therefore, the use of multi-predictors is necessary and therefore the front-up cost of selection. However, this should pay dividends in selecting a more focused workforce, such as in the selection practices of Nissan in the United Kingdom. The quality-oriented organization would ensure that the selection process is managed in a similar way as the shop-floor production line, and the use of specific quality management techniques should be encouraged.

When a candidate has passed the above tests (if any), they are interviewed. On this point, Makin and Robertson (1986) suggested that 71 per cent of a sample of the Times Top 1000 companies in Britain did not use tests of cognitive ability, and 81 per cent used interviews. This is likely to be a small number of applicants — possibly three to five — who may be interviewed that day or at another time and location. If it is scheduled for that same day, there is likely to be a decision immediately made on the person selected.

The selected will generally be notified by telephone, in the first instance, given the job offer, and asked for their decision. It is here that compensation and benefits would be discussed and negotiated. Consequently, all other candidates would have been notified of their position when the preferred candidate has accepted the job and passed any physical examination. However, many organizations telephone unsuccessful candidates after the preferred candidate has accepted the post, prior to physical examination. This may create problems for the organization if that candidate fails the physical examination.

10.7 Training, education and development of employees

Training and development is a requirement in the quality-oriented organization. It is the human equivalent of equipment maintenance and upgrades. However, this similarity stops when predicted performance of machines through upgrades nears 1.0, whereas predicted performance after training is dependent on many factors, including the context of the working situation found after the training programme.

Training is therefore seen as unpredictable in actual outcome, but if it is carried out consistently, with purpose, and consequently reinforced in the workplace, it is a weapon that forms the basis for continuous improvement. It is this structured approach to training and education that sets the quality-oriented organization apart from others.

Training is seen as the process of developing, changing and reinforcing required job-related behaviours. It is current-job-oriented, but is also used effectively to provide the basis for immediate future job changes. On-the-job training programmes obviate the problems of direct application and feedback, and cultural problems related to misinformation or resistance to new ideas. Quality circles generally use this concept; they train their own colleagues, with support staff providing the necessary expertise, if required. The sort of training requirements may include the direct skills needed to perform the actual job, teamwork, process improvement techniques, quality control methods and problem-solving tools. Much of the training carried out in a quality-oriented organization is towards process requirements and the tools needed to monitor and control them effectively. Actual work task requirements are dwarfed by the operational quality requirements. This indicates the depth of changing job contents of frontline workers today. To ensure that products are manufactured to appropriate design and quality-related specifications, much more automatic equipment is used than in a traditional manufacturing plant. This should not suppose that many production lines are designed to operate without workers. Rather the opposite — workers were needed to monitor automatic operations, as automation itself did not substantially replace workers.

A great deal of training is concerned with support staff of production-line technology and processes. There are therefore different skills and competency mixes than in the traditional system. A development with some quality circles is that all maintenance is carried out by the quality circle members, rather than the use of an outside team of maintenance specialists. This means the upskilling of workers, to, some indicate, the detriment of skilled specialist staff — but quality management is also doing this to management, *per se*.

Various types of training programmes have been developed. These include:

1   *Induction programme* — Where staff are introduced to the organization through developing familiarization with their new job, work unit and the relevant work-related procedures for ongoing maintenance. These generally consist of the introduction work routines and safe working practices. The induction programme is directed to increasing the speed at which the new staff member becomes fully operational.
2   *Technical training programmes* — This includes the development of job-specific skills and knowledge in the methods, processes and techniques associated with their particular trade or vocation and is therefore non-managerial in nature. In the quality-oriented organization, it would include the development of training programmes related to other specialized jobs and the techniques used in quality practices.
3   *On-the-job training* — Again, this is job specific, but is essentially on-line. That is, the individual learns while they work, at the point of skills use.
4   *Management development* — This training is directed at management in order to assist in the development of supervisory technical and interpersonal skills. Popular off-line training programmes include the use of MBAs, classroom simulations, etc.

These various targeted training programmes offer different scopes and contents, that are relevant to the different levels of worker and managerial activity contained in an organization. Their structured application is essential, in order to ensure that organizational changes that create shifts in the skills and knowledge demands of staff are met and through training, effectively implemented.

Training for quality must include human (team building, culture, communication), technical (SPC, benchmarking and process improvement skills) and leadership (management of change, empowerment, coaching and counselling) aspects of organizational operations. Widman (1994) suggests that effective quality training should concentrate on balancing these three aspects and in so doing provide the competitive weapon of the internal organization.

10.8 Performance appraisal

Performance appraisal is the systematic evaluation of an individual's job-related performance outcomes. Performance appraisal can provide positive feedback to workers, can be used to provide a basis for recruitment evaluation and to determine what levels of performance can be reached in each work category.

Methods of performance evaluation include graphics rating scales, behaviourally anchored rating scales, essays, ranking methods and interview through management by objectives (MBO). Performance evaluation should be:

1   Job relevant so that many different singular approaches must be developed for each different job category.
2   Easily used by the workforce and therefore should be easily understood and applied.

10.9 Compensation

Compensation is what workers get in return for performing organizational tasks. In basic economics terms, it is exchanging work through the offering of labour — manual or mental for financial — which could be direct (salary or bonuses) or indirect (holidays and other benefits) or non-financial returns — recognition and self-esteem generated in relation to the position held in the organization. Compensation costs organizations perhaps as much as 80 per cent of their liquid cash assets, especially in service-oriented organizations such as educational institutions. However, although individuals need to feel and understand that the compensation that they receive is equitable, quality-oriented organizations generally have in place team-based compensation deals. There is then a conflict. How do you compensate individuals in a team environment? Do we compensate special skills and competencies and then on group output performance? There is no real answer to these questions, except to develop a balanced and flexible scheme that reflects the developed quality-oriented culture.

How is salary or pay determined? In any organizational structure, compensation is likely to be determined relatively to one of three groups of workers. These are:

1   Same job, in different parts of the organization (individual pay development).
2   Similar job contexts, in the organization (pay-level decision).
3   Different job contexts, in the organization (pay-structure decision).

The major problem that occurs here is the development of an equitable evaluation of inter- and intrajob contents. Different jobs will have different skills and competency requirements and therefore the compensation mix will be different.

The doctrine of 'comparable worth' rears it head. Comparable worth mechanisms attempt to ensure that jobs with comparable skills, competencies, effort and responsibilities should be rewarded equally. This could mean that team members are given equal pay because they may have to carry similar job contents when they carry out their next job rotation.

10.10 Workforce relationships and HRM

Much of the pressure of TQM on HRM practices is the need to involve the workforce and get them more committed to the organizational goals. This generally involves the movement from separated structures that are designed to support the individual, to equitable structures that support the team. Consequently, employee involvement programmes have developed as a direct response of this. Quality circles, quality improvement teams, quality of worklife teams are all labels enunciating the same theme, that of teamwork and group orientation. In this regard, HRM and TQM have developed similar veins of operation based on seemingly effective but numerous group dynamics theories.

Caldwell (1984) indicated that Ford's Chief Executive at the time stated 'The magic of employee involvement is that it allows individuals to discover their own potential . . . People develop in themselves pride in workmanship, self-respect, self-reliance and a heightened sense of responsibility'. As discussed in other chapters of this book, employee participation is seen as a major threat, not only to middle management but also to union power. The paradox therefore, is that an individual development process, which seeks to let workers gain, is resisted by the people who represent them. This controversy surrounding these programmes is a major issue that has yet to be resolved effectively. Western organizations essentially see these practices as an attempt at Japanization of Western business culture. Some see this movement to participative structures as leading to more democracy in the workforce and the movement of power downwards through the organization. In this regard, organization development is seen as a practical application of this throughout the last thirty years. So, much of the participative movement as a process is actually reasonably well established. What is different today is that many organizations are flocking to the *new* managerial philosophy of quality, in an attempt to survive, not because top management has seen the light, but in regard to giving workers greater say in their workplace activities.

Participation programmes provide some advantages. These include:

1 Providing a mechanism to develop work-related trust and co-operation.
2 Developing increased commitment to the organizational goals and objectives.
3 Tapping the previously unused skills and competencies that provide the basis for innovation and creativity.

4 Reinforcing the quality-oriented culture, providing tools and developing processes that are quality driven.

Participation programmes may be designed and implemented structurally overnight, but their effectiveness takes time and real committed effort by management and workers. A major problem is that some individuals feel more comfortable by being told what to do, rather than having to solve problems themselves. This is a great resister in itself to participative programmes and rests squarely on the conditioning effects of prior learning and earlier-rewarded behaviour.

10.11 Quality circles

The icon of group-focused quality-related activities in organizations is the quality circle. A quality circle is a group of individuals (up to eight generally) who work on similar tasks, who meet about once per week/month (outside work hours) and discuss work-related problems and help each other develop solutions to those identified problems. The integrative nature of quality circles increase as the perceived dependency of the similar tasks increases. That is, if the circle uses the job rotation method, then the problems of one job for one person also become the problems of any other person in the rotational operation.

However, all quality circles are voluntary and this in itself may provide for changes in attitude of the circle members. Individuals therefore become more committed. One of the disadvantages of quality circles is their, rather, limited scope of action. That is, solutions are not actually implemented by them. Their main facet of operation is to problem solve and in this regard, pass the problem and the developed solution (possibly) over to management.

Several conditions must be in place before a quality circle can be successfully implemented. These include:

1 All circle members must believe in and accept the application of problem-solving techniques to their work processes.
2 Circle members must have knowledge and experience of the process and tasks they have to perform.
3 Circle members must want to work together effectively.

The concept of quality circles suggests that it is a process, a state of culture and therefore a state of mind. These circles offer a powerful way of engendering a sense of ownership, an identity and pride in the output of the circle. To this end, performance is perhaps an outcome that rests squarely on the development of interpersonal relationships and knowledge about the problems of the working processes that the circle members have control over. Quality circles therefore provide a platform for the development of a very effective operational improvement effort.

One of the main preoccupations of quality circles is the need to train and educate its members. Initially at the onset of the quality circle, this may be given by outside training or quality specialists. As time goes on, the circle takes over these activities, perhaps to the extent that outside help is required. Quality circles train their members in the tools of quality, mainly the old tools. As the quality circle becomes more successful, it may also be given the authority and responsibility to carry out its own maintenance of machines and equipment (after specialized training) and the right to choose and select new members.

An *extension* of quality circles seems to be the development of the self-managing team. Here, Sims and Dean (1985) indicate that these could include multi-disciplinary teams that focus on many of the elements considered outside quality circles. Their focus is therefore more wide-ranging, in that the team actually implements the solutions developed to identified problems, which may cover a number of quality circles.

10.12 Chapter review

Human resource management (HRM) is the process of designing workforce measures and activities to enhance the efficiency and effectiveness of organizational performance. HRM is seen as a different, but similar approach to TQM, a key area being the management of the human element. HRM planning has therefore become a major issue because it affects how the organization responds to changing customer, industry, legal and competitive requirements.

Planning for human resources is more difficult than planning for materials or machines. Consequently, greater effort must be attached to HR planning. Various components of HRM planning include internal — needs and availability of staff, machines and processes (present and future) and external situational analysis — demographic factors, SWOT and other market factors consistent with strategic planning.

Recruitment is the process of promoting to and engaging the most appropriate person (with the abilities, competencies and attitude) to carry out the derived organizational activities contained in a job or post. HR planners must therefore have intimate knowledge of job design, the operational requirements of the organization, the direction management is taking now and in the future and its goals and strategies.

Selection is the process of choosing one or more candidates that best meet the stated job or post requirements. The process includes screening applicants through an evaluation of the application blank, background references and pre-interviewing; testing applicants through employment-related and psychological tests; post-test interviews and finally a conditional offer. This process can singularly be of great benefit to an organization or it can hinder the organizational pursuits by selecting personnel that do not match the organizational requirements.

Training and development is a necessary and ongoing requirement in the quality-oriented organization. Training is the process of developing, changing and reinforcing job-related behaviours. If training is carried out consistently, with purpose, and consequently reinforced in the workplace by being relevant, accurate and up to date, it is a weapon that forms the basis for continuous improvement.

Performance appraisal is the systematic evaluation of an individual's job-related performance outcomes. Methods of performance evaluation include graphics rating scales, behaviourally anchored rating scales, ranking methods and MBO.

Compensation is what workers receive for performing organizational tasks. It can take the form of financial — salary or bonus, or non-financial — recognition and self-esteem.

Salary or pay levels are likely to be determined in comparison to workers in the same job in different parts of the organization, in similar job contexts (pay-level decision) or in different job contexts (pay-structure decision). The major problem of compensation is the delivery of a seemingly fair and equitable pay structure.

Participation programmes form the backbone of TQM. Consequently, effective participation programmes are required in the quality-oriented organization. Some advantages include the development of trust and co-operation, increased commitment to organizational goals and objectives and the reinforcing of the quality culture. Part of the development of participation in quality-oriented organization is the development of the quality circle, which is a voluntary, problem-oriented group of individuals who seek to develop solutions to processes they control. In many organizations, quality circles are not encouraged because of their perceived negative impact on middle management and on unions.

10.13 Chapter questions

1. Define the term HRM.
2. Evaluate the relationship between HRM and TQM.
3. Describe the process of HRM planning.
4. Explain recruitment and selection processes and their importance in a quality-oriented organization.
5. Evaluate the impact on an organization of training, education and development issues.
6. Explain performance appraisal.
7. Discuss the basis and influence of compensation issues.
8. Describe the impacts of workforce relationships and HRM.
9. Outline the benefits and perceived limitations of using quality circles in an organization.

CHAPTER 11

Culture and change management

CHAPTER OBJECTIVES

Define the terms culture and change
Evaluate the causes of change in organizations
Determine the best way to deal with change
Determine the best way to implement change
Evaluate the content and effect of resistances to change
Discuss the use and styles of a change agent
Describe and discuss the various ways of implementing change
Discuss the implication for change and the organizational lifecycle
Evaluate change and the need and use of power in organizations

CHAPTER OUTLINE

11.1 The nature of change
    11.1.1 *Introduction*
    11.1.2 *Definitions of change*
    11.1.3 *Causes of change in organizations*
    11.1.4 *What is the best way to deal with change?*
    11.1.5 *What is the best way to implement change?*
    11.1.6 *Resistances to change programmes*
    11.1.7 *Change implementation process*
    11.1.8 *The use of the change agent*
           Change agent styles
    11.1.9 *Process consultation*
    11.1.10 *Change interventions*
    11.1.11 *Change and the organizational lifecycle*
11.2 Culture — group and organizational
    11.2.1 *Definition of culture*
    11.2.2 *Change and the need and use of power in organizations*

11.2.3 *The politics of work relationships*
11.3 Chapter review
11.4 Chapter questions

11.1 The nature of change

11.1.1 *Introduction*

Change is an accepted and normal pattern in today's society, although its affects may or may not be realized or experienced immediately. Nevertheless, change affects every individual directly or indirectly. Significantly, our change patterns and responses to change (if any) are biased to our individual, group and organizational cultures that are developed as we live and develop personally.

Quality management implementation, by its very definition and application, means change in the way things are done, or are expected to be done. It means the negotiation of new agreements for the use of old and/or new skills, procedures, processes, technology, structure and resources. As quality management provides a measurable means for accountability and empowerment, it has major implications for upskilling the base workforce, and the deskilling of management.

11.1.2 *Definitions of change*

Definitions of change are diverse, depending upon the change content and process used. The current understanding of change includes:

> Change involves the crystallisation of new action possibilities (new policies, new behaviours, new patterns, new methodologies, new products, or new market ideas) based on re-conceptualised patterns in the organisation. The architecture of change involves the design and construction of new patterns, or the re-conceptualisation of old ones, to make new, and hopefully more productive, actions possible. (Kanter, 1983)

> In a dynamic environment, change is unavoidable. The pace of change has become so rapid today that it is difficult to adjust or compensate for one change before another is necessary. The technological, social and economic environment is rapidly changing, and an organisation will be able to survive only if it can effectively respond to these changing demands. (Harvey and Brown, 1988)

Change involves evaluating the present and determining a future that is relevant and satisfies present objectives. Change, therefore, involves a vision. Quality planning practices go hand-in-hand with change strategies. Change should be seen in this light as a process to manage what other factors drive. Successful management of the change process will almost certainly mean a successful outcome in regards to the objectives set for the organization, group or individual confronted with the change requirements.

Fullan (1992) discusses the problem of the 'meaning of change'. Here, Marris (1975) cited in Fullan, indicates that in real change there is a 'loss, anxiety and struggle', advocating that change must be painful, if it is real change. Fullan also indicates the differences between imposed 'natural events or deliberate reform' — or voluntarily generated change — 'when we find dissatisfaction, inconsistency or intolerability in our current situation'.

Besides these two elements on the continuum of change (imposed and voluntary), we can also provide some discussion about the subjective–objective reality of change. Here, subjective reality, and the individual's construction of it (Berger and Luckman, 1967) provides the individual's internally isolated world; where covert mechanisms resist or enhance the objective changes occurring *outside* the individual. Real change, according to Berger and Luckman (1967), should therefore be targeted towards this internal reality to ensure effective and long-lasting change results. This suggests why change takes a long time (Kanter, 1983).

Fullan (1992) described the implications to be considered of subjective and objective realities which include the soundness (perceived authenticity of the change), failure of well-intentioned change, guidelines for understanding change dimensions, the reality of the present state, the depth of change and the question of value of a given change programme. Fullan further suggests that 'a change is good depending on one's values (internal reality), whether or not it gets implemented, and with what consequences'. From the subjective point of view, imposed change means the imposition of a subjective reality — someone else's — even if it is cloaked in objectivity. Change implications are considerable.

An individual's perception of reality regarding change may drastically alter the effectiveness of a change programme, because of the inability to determine exactly that individual's perceived reality for each and every step of the implementation process and the eventual outcomes. In this instance, Greenfield (cited in Bush, 1992) indicates that there is a need to 'map people's version of reality'; this is particularly relevant. Mapping an individual's reality, in theory anyway, would provide an objective understanding of where that individual lay in relation to the intended change pattern.

11.1.3 *Causes of change in organizations*

What is causing change in organizations? Since change can be imposed or voluntary (both internally and externally), it is of considerable importance to define exactly what forces are causing the prescribed need for the change programme, and therefore its perceived extent. In the quality-oriented organization, change is the norm. That is, the efforts made to improve work processes and product/service outcomes are designed to be ongoing. In other organizations, this may not be the reality.

Whether change is imposed or voluntary, many forces are derived from outside the organization. A SWOT analysis may prescribe some of the forces that exert this form of pressure (further information on this can be seen in Chapter 10).

A scan of the local newspapers or business journals at any time indicates that organizations are continuously merging, downsizing, upgrading their position in a market by modifying their product, relocating, reacting to changes in laws and political pressure, reacting to technology developments and of course changes in the workplace, through increasing demands to manage multi-cultural workforces. These have occurred through the decades of business in this century. What is perhaps different now is the problem of time. Many of these aspects exact an increasing pressure because of increasing time constraints — time to market of product, time to train a workforce to operate new technology, time needed to react to changes in a market structure (Anon, 1992) for example. Other forces include the reducing distance of foreign markets (in terms of transport/flying times and communication through telecommunications) and the reduction of barriers affording increased competition.

Internal forces of change can come from any of the internal publics in an organization. These include customers and their reactions to product and service supplied (which is why ongoing market research and marketing intelligence are absolutely necessary), workforce and suppliers. They could also include management or marketing consultants engaged for the purpose of developing enhanced managerial strategies in light of, say, a SWOT analysis. Workforce changes may result from increased (or supposedly decreased) performances that result from perceptions of equity and enhanced status (or reduced motivation due to inequities in the reward system, resulting in grievances, turnover and absenteeism).

External forces of change come from any external public. These include politicians, pressure groups, law courts, competitors (products/services), distribution capabilities and in some cases international groups (an international organization will operate in another country's domestic market; consequently, those external forces that apply in the original domestic market may apply in the foreign market). Another external force is the way competitors treat their suppliers and workforce. Benchmarking practices determine the gap and therefore the imbalance between different competitor behaviours.

Irrespective of the forces, managing the reaction to the internal forces is the paramount requirement, because external forces (not all have application) have no meaning unless it is internalized in the organization. Therefore, internal or external forces (normally a combination of these) only create a reaction if management or one of the internal publics seek to implement a change to the status quo.

11.1.4 What is the best way to deal with change?

We can look at change from the point of view of a problem or an opportunity. When change is viewed as a problem, it is generally thought of as being *reactive* to the change force. In this case, being reactive means dealing with change forces on an ad hoc basis. It suggests that management are not actually managing change, but responding to it. Consequently, this type of change management involves little planning or resourcing the change programme and can provoke an incremental change in the organizational direction because of short-termism.

When change is seen as an opportunity, it is generally thought of as *proactive* to the change force. It is opportunity-inducing because management know what is ahead. They are not surprised by developments, because these have generally been considered. In the quality-oriented organization this would mean anticipation of future events and performing to ensure those future events occurred.

Dealing with change depends on management's orientation. Proactive means dealing with anticipated problems, reactive means dealing with problems as they occur. The quality-oriented organization deals with problems before they occur.

Change can be delegated or unilaterally imposed (Greiner, 1967). The unilateral approach, tempting because its implementation is controlled by top management, generally subscribes to the disadvantage of downward-only thinking and practices. The delegated approach, while appealing because of its democratic flavour, may remove the power structure from direct involvement in a process that calls for its strong guidance and support. The paradox is that while change upsets the everyday cultural equilibrium of the organization, pure delegation (abdication?) cannot be enacted unless the whole organization already accepts the behaviours that delegation means, demands and delivers. Consequently, the impact of the change programme will be reduced and the focus of the change programme will be on the content, and not on the process surrounding it.

In the situation where the change programme objectives and the process are totally new to the organization, the actors must first become accustomed to the change programme process before enacting the content of the programme. Here, a unilateral approach would normally be required initially.

Greiner (1967) also suggests the development of a shared power approach. It is presumed that if management are serious about their change programme intentions, then a shared power approach would evolve anyway. The amount of power share would seem to depend upon the criticality of the nature of the change programme and possibility as the criticality rises, the power share diminishes directly, as a consequence. This is somewhat contrasted with Louis and Miles (1990) (cited in Fullan, 1992), where they determined that 'power share' was crucial and necessary — and the results provided an example of what 'power share' can do and what Fullan also suggests that 'implementation is very much a social process'. This cannot really be argued against, since the process of change necessarily involves people.

It therefore becomes another paradox — that of the need to ensure change, in a given form, as opposed to the need to provide mutual developments of change objectives and processes of implementation. Depending on the change programme content, almost all change programmes that are to develop the 'human' aspect of organizations, will promote problem solving rather than power and political development (French and Bell, 1984), which may, however, provide an indirect result of the change programme and by empowerment through the development of skills, training and education.

Collaborative management practices that generate systematic problem solving — normative/re-educative strategy — will provide more reinforcement of the objectives of the change programme than a power/coercive strategy framework, as developed by Chin and Benne (1979).

It can be inferred from above that problems with change could take possibly three forms:

1. The change process itself — The way change is managed.
2. The change results — post-cultural requirements — Changes in expected behaviours.
3. The prechange culture — Behaviours that resist the change process or change outcome.

Highlighted problems of change in these circumstances were:

1. Change is perceived as too fast.
2. Insufficient dialogue — before, during and after change programme implementation.
3. Change programme not related to pregroup norms and cultural expectancies.
4. Lack of change skills exhibited by both management and workers.
5. Change programme has not been completely thought through with regard to implications of change programme outcomes, etc.

11.1.5 *What is the best way to implement change?*

There seems to be no best way to implement change. Change programmes cannot be implemented in a group or departmental vacuum, because the group is seen as an open system (French and Bell, 1984), where interactions occur with other groups or departments within the organization and they need their support — both in terms of resources and commitment. Without these elements, the change programme will **not** succeed.

French and Bell (1984) also suggest that change programmes are interventions that are structured activities directed towards selected target groups. Many problems exist when implementing a change programme. To eradicate such problems during implementation, the change programme should be structured such that:

1. The relevant people are there. This includes all staff likely to be 'affected' by the change (target groups and any other group who will assist in the process of implementation of the change, e.g. the training department).
2. Its aims and objectives are clearly communicated to everyone. There is a need to ensure that the change *message* is seen as authentic and as believable

as possible, is clearly visible, and a trusted icon of senior management — who should have the task and responsibility to carry this initial communication.
3 Positive reinforcement of the change programme is developed in order to provide increased inertia for the continued implementation of the change programme.
4 It contains both experience-based and cognitive-based learning practices.
5 Through human icons, develop the change programme through a process of ensuring top management commitment and rewarding any commitment to this.
6 The change process is developed so that individuals *learn how to learn* (Schein, 1987). Here, workers should be helped to help themselves to become motivated to change their own behaviours, to become innovative and experiment and thus to ensure that their skills and knowledge are effectively developed.

Quality management is not only essentially focused on *changing* individuals — workers and management — but also others, such as suppliers and even the community. Harvey and Brown (1988) discuss a model of environmental adaptive orientation which could be applied to many organizations. Prior to the adoption of quality management practices, top management seem to have a conservative management orientation, coupled with a low adaptive orientation and perceived environmental stability, whereas other more open functioning departments, e.g. the training department, who have the responsibility for developing staff appraisal programmes, usually exhibit a satisficing management orientation, which emphasizes centralized decision making in a stable environment.

Other organizations, perceived to be involved in rapidly changing environments, have a demonstrated anticipative management orientation — a dynamic environment with pressures for high adaption requirements. These different management biases create micropolitical problems (Hoyle in Bush, 1992) and any relevant cultural problems therefrom. This is because, at any one time, organizational groups can differ in practising their management orientation between those two adaptive continuum points. This approach highlights the difficulties that are inherent in a wide-ranging but focused change programme, as quality management programmes usually are.

11.1.6 *Resistances to change programmes*

For some people, change means the development of a stimulating and exciting working environment; for others, it can be both painful and anxiety generating. The development of effective change in many organizations is becoming a necessity for all concerned — whether worker, manager, industrialist, government or community member.

Barriers to effective change includes micropolitics, individual and group cultural dispositions, the predominant organizational model of management and the sense of whether the change is being imposed or is sought voluntarily. The consequences of not managing the change may lie in wasted resources, increased conflict and politics and even increased chaos. The views of a *good* change or a *bad* change may not be apparent from the content of the change programme, but may be seen distinctively when we consider who the change affects and whether they deem the change as positive or negative. Thus change outcomes may be more political than practical, and more social than functional.

Watson (1966) assumes that there will always be resistance to any form of change. It must be emphasized here that it is people who offer resistance to change programmes, not the programmes themselves. Watson suggests that change moves through a lifecycle of five phases, where it seems 'that a battle is being waged between those trying to bring about change and those resisting that change'. Some change programmes may die in the process of moving through these phases, so continuous evaluation of the stage of development is a useful tool to use.

According to Harvey and Brown (1988), resistance to change can be predicted, cannot be repressed in the long run and must be managed. Therefore, the method of implementation must consider this factor. Resistance to change can manifest itself in many forms; however, the perceived power that an individual has may assist or resist the advances of a change programme. These forms include self-orientation, fear, habit and social.

- SELF-ORIENTATION

This suggests that an individual will resist the outcomes of the change programme because of the perceived negative affect it has. For example, many change programmes require a restructuring of the organization or part of that organization and the patterns of power may change as a consequence.

- FEAR

All change introduces the unknown to an individual. Consequently, blind fear may be the main resister to the change programme. Communication will provide a basis for reducing this aspect — communicating the outcomes of the change programme, the process of delivery and the positive aspects, such as new experiences and new skills. Overcoming fear can mean a prime resister has been nullified.

- HABIT

Individuals who have worked the same way for years resist change, because they have forgotten how to learn. They are comfortable with the status quo and know the lines of communication and power in an organization. Change means that an individual's work habits must also change. This brings in fear and its consequences, as described above. The exact nature of the change programme will determine how much of an individual's work habits are likely to be affected. Total change — as in separation through job loss — is likely to be resisted greatly. Less change will likely meet a reduced amount of resistance. The culture of a group that is expected to move from a work pattern of certainty to one of developing flexible working practices can leave some individuals totally out in the cold. Reassurance, training, education and support may be necessary in some cases.

- SOCIAL

Opinion leaders in groups apply pressure on group members to conform. Consequently, their perceived view of the change programme is an important view. Where these opinion leaders personally resist the change programme, then it is also considered so will other members of that group. This suggests that managing opinion leaders is as important as managing the group as a whole. Targeting communication efforts at these leaders will provide a more positive means to nullify resisters.

Reducing resistance to change programmes involves the management of at least four resistance forms. As a basis for this, the effective management of communication, education and training practices, methods of negotiation and conflict management, and power are necessary. The degree of application though, depends on the circumstances facing the organization.

A major consideration when evaluating change in an organization is the degree of perceived change, as opposed to the actual change created or intended. The change implications model contained in Figure 11.1 indicates the effect of this on resistance.

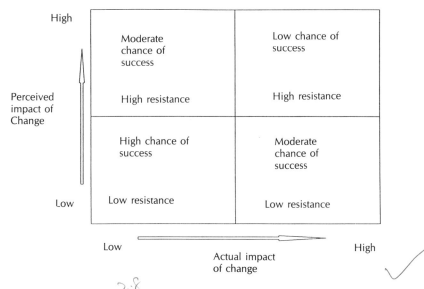

Figure 11.1 The change implications model.

11.7 Change implementation process

Fullan (1992) argues that there are six themes in the implementation process and that these consist of vision building (providing direction), evolutionary planning (adaption of plans to fit the present realities and conditions), monitoring/problem solving (monitoring allows tracking of change progress towards the vision and thus provides a platform for problem solving in order to get back on track), restructuring (focusing on the outcomes of the change programme), staff development/resource assistance (change needs support by management and resources to visibly provide management commitment), and initiative taking and empowerment (power share and supported and stimulated initiative taking).

The process of implementation could be through top-down, where a top level change task force analyzes how to alter strategy and structure, recommends a course of action, and then moves quickly to implement it, or bottom-up, where change is perceived as much more gradual, and the change task force consults with managers at all levels in the organization. The emphasis would be on participation and keeping people informed about the situation, so that much uncertainty is minimized (Beer, 1980).

The impact of the proposed change programme on the operating practices of the workers can be extensive. It would seem that many people are waiting for this type of programme in order to update their skills and competencies, so that many workers do not fear its implementation.

Conversely, many administrative staff, especially in staff departments, fear that the change programme will actually make their inadequacies much more visible and thus want to resist the change implementation. Top management are seen to have more power than any other group in an organization, and control resources. The more positive top management is to the change programme outcomes, the greater the likelihood of its successful implementation.

A framework for the implementation of a change programme can be seen in Figure 11.2.

Within the confines of the model, the implementation process provides a number of stages that include awareness, diagnosis, application and review of the change programme.

Lewin (1947) introduced the concept of there being three stages in the change process. These are the unfreezing, changing and refreezing processes. They are used to break-up the change implementation process in such as way as to indicate that different approaches are required in the three different stages:

1. The *unfreezing state* refers to motivating the concerned personnel to accept the change programme possibilities, and therefore the change outcomes, by understanding the need for change or that the present situation is inadequate.
2. The *changing* state refers to attempting to motivate individuals to experiment with new behaviours. This in turn is expected to create changes

Identification and evaluation of change forces

Diagnosis of implications of change forces

Development of change strategies and problem evaluation

⇩

Selection of change and implementation method(s)

⇩

Implement change programme

Evaluate the effectiveness of the change programme

Figure 11.2 Change implementation framework.

in attitude that will permanently make further changes to behaviour, hopefully in the anticipated direction.

3   The *refreezing* refers to anchoring an individual's behaviour at the point where no further change is required. This is rather difficult to do in practice, and indicates that once change is set in motion, it is difficult to control. Given that the environment of a quality-oriented organization is almost perpetually in a state of change, then control by managers is impossible, it relies on the workforce to take those control measures themselves.

Another diagnosis method that is deemed very useful is Lewin's (1951) Force-Field Analysis, which was first developed in 1938. This method is essentially used to determine the forces for and against a particular change programme. The method is illustrated in Figure 11.3.

The model indicates that in an organization there are driving forces for the change programme and resister forces against the change programme. Driving forces are forces that the change programme has on its side. Consequently, reducing the resister forces will move the change programme from its original position to some newly defined position. However, determining exactly that position is difficult, even if supposedly objective parameters, such as absentee or accident rates, are used. The model accounts for the perceived degree of force for or against the change programme, which is illustrated by varying the length of the arrows, i.e. the longer the arrow, the greater the perceived force.

However, in practical applications it is difficult to determine what are the causes of the perceived resistance and the actual status of that resistance over time. The model seems incapable of mapping the generated changes over time, as it can only be used to determine a subjectively personal snapshot. It is therefore useful for planning, but not for tracking.

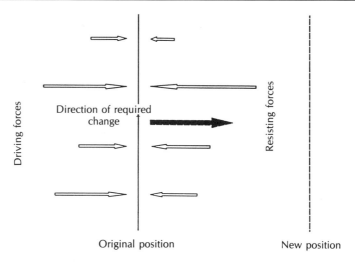

Figure 11.3 Lewin's force-field analysis.

11.1.8 *The use of the change agent*

A change agent is an individual — normally independent of the outcomes of the change programme — who offers objective advice about the planning, development, implementation and review of a change process. A paramount requirement is the development of a good working relationship between the change agent and the client (for our purposes, the individual responsible for the development, application and evaluation of a change programme). However, determining exactly who the client actually is can be difficult. For example, is the client the top manager who engages the change agent or is the client the manager of the department that the change agent is assisting? If the change agent represents top management, then the change programme takes on a feeling of being *imposed*. If the change agent represents the manager of the department, then there can be increased ownership of the change programme by the department members. In this circumstance it is seen as *voluntary*. These are problems that have to be managed effectively, and are part of the political fabric of the organization.

An *external* change agent is an individual or group who does not rely in any way on the client system. In this way independence is usually assured. Generally, the external change agent is invited to help the client system. Internal politics and power are short-circuited, as every individual is seen on an individual level. A disadvantage of the external change agent is that they tend to be unfamiliar with the everyday operation of the client system, have no knowledge of the patterns of communication or the pervading culture. In such circumstances, the change agent relies on the contact person's power base in order to unlock vital information.

An *internal* change agent recognizes and knows the power system, lines of communication and possibly the pervading culture of the client system. The choice of the internal change agent may determine how others feel about the change programme, i.e. if the individual is a trusted member of management and has a successful track record of implementation, then the client system will more likely believe the change message. However, since the change agent is considered independent for the purposes of the change programme, he/she may still act in similar terms to the external change agent.

But, when the actual change programme creates severe organizational problems, the independence of the change agent is compromised, as the change agent is seen as representative of a group in the organization. Consequently, rather than enhancing the change process, the opposite often happens. Internal change agents seem to provide the organization with a means for helping itself manage change, and if the change programme is planned and implemented effectively with sufficient commitment, then the internal change agent is a very useful addition to the skills armoury of the organization.

Change agent styles

Harvey and Brown (1988) suggest that there are five different styles that a change agent can use in the process of managing change. These styles are synonymous of Grid OD (Organization Development) as developed by Blake and Mouton (1964) where the two dimensions used are concern for effectiveness and concern for morale. The four styles, that are deemed to be appropriate for our discussion, are:

1. *Agreeable* — This style suggests a change agent behaviour that wants to keep the situation on an even keel. Essentially, the requirement is to get along as amicably as possible. However, the result is an ineffective change programme, because no one is convinced of the change message.

    Who is going to commit themselves to painful change when the individual assisting the change process does not believe in it either? This style is also projected from the position of low power (from the change agent's point of view) and influence. This can occur when a trainee manager or student on placement is given the job of assisting the implementation of the change programme.

2. *Analytical* — This style places emphasis on the functional areas of the change process. Little regard is taken of the socio-human aspects of the change programme. It reflects the objective or expert-oriented change agent who sees the change programme implementation as more of a structural change, rather than requiring the change of individual attitudes. This style is generally used for change programmes that have little depth, in that the change is rather superficial in terms of perceived human consequences.

3. *Supportive* — This style is diametrically opposite that of the analytical style.

The emphasis is placed on the socio-human context of the change programme. It reflects a change agent's lack of technical ability or the diagnosed need to concentrate on the socio-human aspect.

The style is useful for intergroup interventions, where the development of good group process is paramount to any structural changes. Much of the effectiveness of a change programme, in terms of the technical aspects, may be compromised, as the change programme focus is moved.

4   *Integrative* — This style directs energies towards recognizing and equitably balancing a need for effectiveness and socio-human requirements. In this way, the integrative practices of the change agent style consider communication, group member roles and functions, group problem solving, power and conflict solution.

The styles a change agent uses can be a reflection of the organizational culture and climate or it may be a reflection of the change agent. Consequently, in order for the change programme to be successful, it requires that the change agent uses an appropriate style that matches the organization environment. However, as change programme objectives turn into reality, especially if the change programme is wide in scope and depth, then the style of the change agent may be required to change too. The change agent therefore must be dynamically tuned to the change programme and the reactions and changes of the client system.

11.1.9 *Process consultation*

This model, developed by Schein (1987), is applicable to all interactions between a manager and subordinate, the manager and change agent and the subordinate and change agent. The theory is based on the assumption that managers and change agents act in a consultative capacity in order to help others perform. In regards to change, the change agent primarily helps the organization learn for itself, otherwise when a change programme needs to be implemented again, the manager has no option but to call in the change agent.

Process consultation (referring to how things are done, rather than what is done) is essentially about the application of a helping managerial process that seeks to ensure that individuals get the right kind of assistance in order to solve the problems at hand. It is a 'set of activities on the part of the consultant that help the client to perceive, understand, and act upon the process events that occur in the client's environment' (Schein, 1987).

Schein indicates that there are three models of consultation — the doctor–patient model, the expert model, and the process consultation model. Each applies in different situations that can be determined during the diagnostic stage. Figure 11.4 illustrates the process consultation continuum.

Each model has some basic assumptions that need to be understood in order to evaluate the models effectively. These assumptions direct the change agent to apply the most appropriate process model to the circumstances diagnosed.

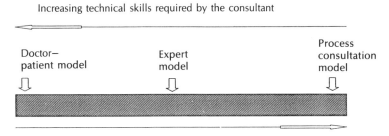

Figure 11.4 Process consultation continuum.

Even in the diagnosis stage, a change is expected to lead to the application of one of the models as the engagement of dialogue between the manager and the change agent requires this.

The *expert model* is based on the premise that the client has successfully diagnosed the problem, matched the consultant's capabilities with the identified problem, effectively communicated the problem to the consultant and has considered the implications of receiving the reciprocal expertise. The use of this model is only advised when the client understands that much of the success will rest on their capability of diagnosis, rather than just the consultant's. This does not move the right of the consultant to reconfirm the client's diagnosis, but does indicate that the consultant somewhat relies on the problem advised. If a client cannot determine the real problem, then the client will consider which of the next two consultation models to use. A major issue here is that the client passes the problem over to the consultant, that is, problem ownership transfers to the consultant.

The *doctor–patient model* is based on the premise that the client knows it is hurting somewhere, and has a vague idea about the cause. It is synonymous with the doctor and patient consultation where the patient has a pain, and the doctor prescribes a pill in order to take the pain away. The consultant's role is to determine a diagnosis and to prescribe a remedy. Whether the client takes the remedy as prescribed is left to the client. Consequently, failures can be attributed by the client to a wrongly diagnosed problem and remedy, rather than because the client mistook a symptom for a cause. This means that organizations using this model of consultation must be very sure of the problem at hand and to be serious about taking the remedy offered — whatever that was. A major issue here is that the client passes the problem to the consultant, that is, problem ownership transfers to the consultant.

The *process consultation model* is based on the premise that ownership of the problem rests squarely on the client system, not the consultant, throughout the consultation process. Another major theme is that the consultant helps the client system to learn for itself, helps it to problem solve, and helps it to manage

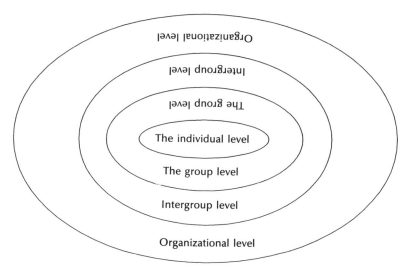

Figure 11.5 Change interventions.

its system appropriately to effect a successful problem solution. The consultant who uses this model does not have to be trained in the technical nature of the change programme, i.e. is a technical expert, but does have to be trained in developing good change process. This model treats diagnosis and intervention as an integrated process, rather than one that can be separated, as in the previous two models. This model therefore, emphasizes the learning needs of the client system, rather than the consultant.

11.1.10 *Change interventions*

This suggests that interventions are planned, systematic processes that are designed to solve specific problems, realizing opportunities and reducing ineffectiveness in an organization. Change interventions can be targeted to four distinct but related areas. These are contained in Figure 11.5.

Different change interventions are more effective at one level than another level (individual, group or organizational). Consequently, using the most appropriate change intervention will ensure that the change programme objectives are met. There will be times of course, when more than one intervention can be applied. In these circumstances, the experience of the change agent will help to determine the method that is most appropriate. The change interventions relative to level can be seen in Figure 11.6.

Interventions can be seen as an interaction between the client system requirements, the change agent and the change strategies adopted to provide the necessary and successful change outcomes.

Personal	Team	Intergroup	Organizational-wide
Job design	Team building	Process consultation	MBO
Job enrichment	Process	Grid OD	Survey feedback
Goal setting	Job design	Intergroup development	Action research
Managerial grid	Quality circles		Goal setting
	Role analysis		
	Grid OD		

Figure 11.6 Change interventions and organizational leadership.

11.1.11 *Change and the organizational lifecycle*

Greiner (1972) suggested that there were four stages in the lifecycle of an organizational. An adapted version is illustrated in Table 11.1.

Here, the demise stage is added, as organizations tend to die if the adopted strategic strategy does not match the constraints of the market, i.e. if the strategy does not provide a basis for keeping or increasing market share or satisfying customers in both the short and long term.

The evolution through the various stages depends on the size of the organization and its developed approaches to internal and external pressures during each stage. It would seem that each stage must be mastered in order to progress to the next stage, and that the experiences of the previous stage provide a basis for the organization to achieve its goals in the next stage. However, demise can occur at any stage in the lifecycle.

11.2 Culture — group and organizational

11.2.1 *Definition of culture*

All groups adopt their own individual culture and therefore seem unique. During group development the group's culture changes, but this still provides the group's unique cultural signature tune. Each group will be different to every other group formed or otherwise at the same time, with the same number of members, from the same community or ethnic area or work population.

Table 11.1 Lifecycle stages

Stage 1	Entrepreneurial
Stage 2	Collectivity
Stage 3	Formalization
Stage 4	Elaboration of structure
Stage 5	Demise

But what is culture? The meanings adopted by management theorists are generally developed from social and anthropological constructions and ideas surrounding cultures. The culture of a group is dependent upon many variables and include (Harvey and Brown, 1988) language, dress, patterns of behaviour, value system, feelings, attitudes, interaction and the developed group norms of the members.

Hofstede (1984) suggests culture is 'the collective programming of the mind which distinguishes the members of one group or society from those of another'. This concept could be taken further by indicating that individuals are themselves self-contained socio-units and therefore are distinguished by their own cultures. Bion (1959) however, determines that group culture arises from conflict between the individual's desires and the developing group mentality. This is brought about by the relevance of the type of task, its adopted structure and the leadership-style offering.

The literature contains many sources of investigations into the effects of culture on management (Weinshall, 1977) but these are at the macro level and not at the micro or organization subunit level. This tends to be frustrating, because it is small groups who do the work in all organizations, irrespective of the size of the organization.

Skinner (1969) commented, in behaviourist terms, 'a culture is not the behaviour of people "living in it"; it is the "it" in which they live — contingencies of social reinforcement which generate and sustain their behaviour'.

Norms can provide an indication of the culture of a group (Schein, 1985). Norms are an outward emanation of enforced behaviours that are viewed as important by the majority of group members. Also, the management style and the set of norms, values and beliefs of the group's members combine to form the group culture. Harvey and Brown (1988) suggest that the group culture must

Table 11.2 Political mechanisms

Activity area	Political mechanisms			Symbolism
	Resources	Power groups	Subsystems	
Building the power base	Control of resources Acquisition of other resources	Sponsorship or support by other power groups	Alliance building Team building	Building on legitimacy
Overcoming resistance	Withdrawal of resources	Breakdown of power groups	Foster momentum for change	Attack or remove legitimacy Develop confusion and conflict
Achieving compliance	Giving resources and information	Removal of resistant power groups	Participation	Rewards and assurance

Printed with permission from Johnson and Scholes, *Exploring Corporate Strategy*, Prentice Hall, 1993 achieve goals as well as satisfy the needs of members in order for the group to be effective. This is consistent with the team building strategies found in the application of TQM initiatives.

Henry Ford once said 'You can take my factories, burn up my buildings, but give me my people and I'll build the businesses right back again'. Ford was talking about the culture of his business.

Culture, therefore, is not only vested in machines, equipment and other physical entities but more importantly is the result of the subjective minds of each individual in any given group. There are major problems with group cultural development. Group cultures differ inherently, but there is still a core of characteristics that link some similar culture types. These cultures evolve, transform or adapt to a changing environment to survive. Culture transforms individual behaviours, is unique to each group, and develops internally.

11.2.2 *Change and the need and use of power in organizations*

Politics, in its broadest sense, is the operation of power. It embodies the mobilization of resources towards a given end, and the threat and/or use and implementation of sanctions. The concept of power seems to involve immense problems in its definition and theory of application in organizations, and none more so than in the quality-oriented organization.

Quality management can be viewed as necessary for modern organizational requirements of survival and competitiveness. This requires someone to nurture, develop, shape and push an idea, notion or philosophy until it takes useable form as, say, a quality management system. The individual does this through the use of power. The search for innovation (quality change programmes) in an organization consists of a search for power.

This power can be seen as being derived from three areas — information, resources and support. It is the continuous move towards gaining control of those areas that gives rise to conflict in groups. It is not fully understood that this search and use of power can have two consequences, first, that it adds weight to the innovation or initiated quality change programme and second, that it develops resisters to the quality change programme.

The problem of change as part of a continuous strategic activity has essentially two levels of implementation. One, organizational, where departmental change occurs, the other at operational level, where the work group change occurs. Management need to realize and understand that management activities can take on political dimensions. Therefore, the importance of understanding the political system contained within an organization is paramount to the successful planning, implementation and reinforcement of the change programme. Table 11.2 (adapted from Johnson and Scholes, 1993) summarizes some of the political mechanisms contained within organizations.

These include resources, power groups, subsystems and symbolism; other

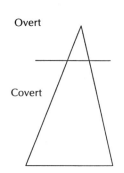

Overt	Pressure groups, power groups, technology Supporters' goals and Strategies Policies and procedures
Covert	Political — agendas — strategies — norms — culture Feelings, values, beliefs, assumptions

Figure 11.7 Political iceberg.

mechanisms exist, but the discussion will be limited in the context of this chapter. The control and manipulation of organizational resources — human, physical and financial — is a source of power, e.g. acquiring additional resources or being identified with important resource areas or areas of expertise; the ability to withdraw or allocate resources can be valuable in overcoming resistance or persuading others to accept change; the careful use of information or news to counter that being used to justify opposition to change may be seen to be important. The very essence of managerial and therefore political activity means that demands will be made by different groups for similar or very different reasons. These demands could be put 'by virtue of claims recognised in our society as valid on legal grounds or on economic or other grounds which are socially approved' (Burns and Stalker, 1961). If the organization and its political diversity is a 'balance of competing pressures from each group' (Burns and Stalker, 1961), then it could be determined that there is some sort of political iceberg, as seen in Figure 11.7.

The political iceberg consists of overt and covert political mechanisms, where both are seen to influence the organization towards its goals for efficiency and effectiveness, but not in equal amounts. The overt and covert mechanisms can be applied equally to individuals and groups and organizations. The operation of the political iceberg, as far as groups are concerned, is with strategic empowerment to ensure survival and competitiveness.

Handling power groups can be of crucial importance with the design, development, implementation and review of the success of quality change programmes. Association or support from power groups can help build a power base. It can also help in overcoming resistance to change. The breaking down or dismantling of power groups provides another means of overcoming the resistance posed. This gives us some sort of paradox: you may need to use or build power groups to create support, on the one hand, but you may need to dismantle them to reduce their perceived power base on the other. Consequently, the success or failure of a change programme can be based entirely on political overtones and the application of power, rather than the efficacy of the change content. Groups do not share interests simply because they are related in the division of labour. This has implications for power groups, such as quality circles, middle or even top management.

The reality in industry, in Britain at least, is that due to the immature development stage of quality circles, competition still abounds between these types of groups. The use of quality circles by McDonald's, for instance, is seen by the workers to increase efficiency, not to provide better working environments for the workers, as they were led to believe. In essence, the guise of 'quality circles' has not effected the empowerment sought by workers and may even have created further resistance to future change programmes, which will need to be very carefully handled to ensure successful implementation. The gaining of recognition, one of the facets of quality circles — and also one of Hertzberg's motivators — can itself be a destructive force as different quality circles become more politically radical by using that recognition to make further gains for the circle members — an independent, but powerful means to use change programmes as political platforms to change power structures, rather than to just to effect change in work practices, for instance.

Kast and Rosenzweig (1979) define a system as 'an organised, unitary whole composed of two or more interdependent parts, components, or subsystems, and delineated by identifiable boundaries from its environmental supra-system'. Thus, a system denotes some form of interdependency or interaction of its component parts. This can demonstrate that, although groups compete for power and use the platform of information, resources, and support as a means to increase or decrease perceived power, they are by definition still interdependent. Building alliances and networks may result in gaining assistance in overcoming resistance to change programmes, especially with larger groups.

Effective subsystem management occurs when *all* of the organization is converted to the acceptance of the change programme. Until this occurs, the danger remains of power groups wielding claim to their 'turf' and resisting the change programme. Likewise, the power groups that are sympathetic to the change programme need to be rewarded publically — and are therefore visible — to provide enhancement, motivation and recognition for their positive support.

Politically this can take various forms such as:

1   To develop, test and reinforce the quality change initiative.
2   Power building and sharing — power-base building requires considerable management actions, and because of the required interaction between existing and potential power groups.
3   Overcoming resistance — attempting the change programme from a lower perceived power is almost bound to failure.

11.2.3 *The politics of work relationships*

The aspect of reward negates some of the micropolitical — 'the strategies by which individuals and groups in organisational contexts seek to use their resources of authority and influence to further their interests' — problems espoused by Hoyle, in Bush (1992) where Hoyle suggests that 'the head, who has a high degree of authority, and can exert a considerable degree of control over organisational activities, will also have at his disposal a wide range of micropolitical strategies'. Since the rewards are visible outcomes of change effort, then covert acts of micropolitics will be reduced but not entirely eradicated, because differences will always be present between groups and groups will always be trying to get a better deal. However, overt and covert acts by a given group are designed to ensure they get what they want, and more.

Skilled workers possess labour power of a specialized character, thus rendering management more than ordinarily dependent upon co-operation (Sabel, 1982). The labour process therefore must be managed to accept this as a basis. The skills employees gain may support powerful resistance to various forms of change programmes, especially in the area of the *knowledge* professions. The paradox is that quality change programmes seek to **increase** employee skills, not decease them, and thus increase their inherent power at the expense of management. This manifests itself when whole layers of middle management are removed from the hierarchy, giving legitimacy to the increasing power build of the lower-level workers.

Ironically unions, generally accepted as bureaucratic power cultures, resist this move of empowerment, not because of what the quality change programme can do for the workers, but what it does to the union. With increased skills, pay and greater personal development prospects, workers have a much reduced need for unionism. The union's traditional role of 'fighting' for the maintainers (in Hertzberg's terms, e.g. conditions, pay, etc.) and very little for the motivators (e.g. responsibility, recognition, etc.) have only provided workers hungry for the motivators, because conditions, pay, etc., have become mechanistic in nature and outcome. Workers have therefore become a different power group from the union, with a different emphasis and objectives that need satisfying. It is the secular build of individual skills development and its demonstrated achievement in practice that provide a powerful and visible basis for positively reinforcing the change initiatives. However, the quality change programme generally brings the tools, education and training and other processes that could be ascribed within a quality management system.

The use of a bureaucratizing quality management system unfolds on the worker, not as a means to have freedom over one's job, but as a means by which management are more effectively informed about individual worker performance than before. This Tayloristic comparison, made with the introduction of mass production, can certainly be applied to the introduction of a quality system and the information and tagging that can occur for each and every error and defect produced. It would seem that George Orwell may have been right about one thing, the increase of systems that allow more complete surveillance of performance of personnel in organizations.

Since politics is the operation of power, it is also about the operation of control. It is the level of influence of this power that can help or hinder individuals and groups gain or lose control of a given situation. If the level of influence is low, then the level of power is low and consequently the level of controlling that given situation is low too.

It is submitted that, in behaviourist terms, the power of influence rests, not with the party seeking to influence, but with the party who perceives the level of influence. This is a rather fundamental point, because if a party seeking to influence another party has not carried out sufficient 'external' research, then the limited information about the relative power of a given group may be wrong and the influencing strategy will fail. This neatly brings us back to the much offered maxim that *information is power*.

Political arguments about the influence of groups are greatly enmeshed within the development of group norms, determined as a covert group activity (see Figure 11.7), because it is part of the culture of the group. In the Hawthorne studies, the peer pressure of the 'wiring loom' group resulted in group control of the individual work performance, a case of the politics of the group influencing the politics of the individual.

Outside political influences (political parties, unions, management and workers, etc.) can have a bearing on the inner political machinations between groups within the organization. When society expects that each individual consumer is treated as a single market for the organizational wares, then the influence on the organization can be just as severe as the internal political dialogue. Indeed the very nature of external political influences can be viewed as the trigger for internal political activities, because every single individual in an organization is also a member of society. A similar political operation of power exists outside organizations, because an organization is one of many groups with vastly differing political agenda, resources, goals, strategies and motivations.

Let us turn our discussion to the internal politics of a single group, rather than the politics involving more than one group, as the discussion would not be complete without touching on the significance of intragroup politics. The ultimate the group can achieve, in terms of group process, is to carry out individual work activities towards the goals of the group, even at the expense of subordinating individual goals. This sacrifice individuals make is to ensure group dominance and group survival. Does this really happen in practice? If not, what prevents total group dominance? To answer this we need to examine what sort of power bases individuals and groups possess. French and Raven (1959) introduced the notion of five interpersonal power bases — legitimate, reward, coercive, expert and referent. The group is deemed to exercise the first three, whereas the latter two are exercised solely by the individual and are based on personal qualities. The effect is that individuals, having different abilities to express their 'power', enact politics differently within any given group.

Kanter (1983) discusses the 'myth of the team'. The mythology that surrounds a team suggests that when a group is a team, no differences exist between its members. Each individual is an actor who shares equally in its operations and members pretend that other individuals are not more able, or that there are individuals who are not dominating the group practices and influencing its outcomes. This directly provides for pluralistic ignorance — where every group member knows individually about the differences, but assumes no one else does. This particularly occurs during the start-up of groups such as quality circles. 'People with less to contribute because they are less informed do not feel comfortable seeking help in getting more information to contribute more because of the myth that everyone has an equal chance to contribute' (French and Raven, 1959). Contribution in this sense does not mean real participation because the power differences prevent a contribution, which in itself perpetuates the power differences as time goes on and the group develops.

While groups wallow in pluralistic ignorance in the early stages of development, dominant individuals are busy creating the group norms that will be difficult to change as the group develops. Individuals, through their political games, controlling or attempting to control information, resources and gaining or withdrawing support, 'guide' the group towards a group culture that is relevant to the dominant individual requirements. To compound issues, due to pressures of autopiotic systems, dominating individuals would seem to try to use their political power to shape a group in the same manner as his/her *mother* system, another group or the organization itself. If the organization has a track record for successfully implementing change, then this is likely to support the development of the new group and provide reinforcement for its outcomes, that of implementing the change programme.

11.3 Chapter review

Quality management, *per se*, means change. Change means anything that upsets the status quo or equilibrium of a work situation. It includes the development of training and education mechanisms in response to new technologies, changing customer requirements and management practices.

There is no one best way to implement change. Organizational-wide change programmes are effective, but take a long time in implementing, as well as providing many problems in relation to control. More focused change programmes are useful, but may suffer from a change vacuum, where external relationships may be affected. Structured approaches will mean that the relevant people are informed of the changes and how it would affect them. They would also be given information on the objectives of the programme, how conflict will be dealt with and what resources and support are available to help individuals manage the change process effectively. Themes involved in the successful implementation of a change programme include vision building, adaption planning and continuous monitoring of change progress.

Other factors that may need to be considered are the effects of micropolitics and power through resource use and control.

Culture has many meanings, and includes the set of similar behaviours and attitudes that set one group apart from another. The culture of a group, department or organization can assist or inhibit change. Therefore, effectively managing change is a key factor in managing quality in an organization.

Politics is present in all groups, and therefore organizations. Politics comprises those activities that are used to acquire, develop and use power and other resources to obtain one's preferred outcome, when there is uncertainty or disagreement about choices. Political influence has both an external and internal implication and the trigger for political behaviour can come from either. Group or individual politics has implications for change, especially when a group perceives that acceptance of a change initiative could mean their demise. Various political mechanisms that are deemed to exist help or hinder any change initiative, and that initiative must be carefully managed to ensure that it does not backfire, with a subsequent and uncontrolled initiative, under their terms and in their direction. The use of the political iceberg illustrates that the 'power' of politics is essentially driven by the covert mechanisms that bear on the ability, skills, resources, coalitions and information control of the group in comparison with other groups within the organization. Most successful groups are successful because they understand how the 'system' works and are willing and able to manipulate rules and procedures to mobilize support and overcome opposition so that the change initiatives are accepted and the objectives of that change programme are met.

Thus, politics through the operation of power, should not be construed as something negative, but merely as a skilful means of getting things done (change programmes implemented) in a way that may or may not reflect purely rational decision making and a totally objective consideration of all the 'facts' available.

Politics is about people and people are politicians whether they like it or not.

11.4 Chapter questions

1 Define the terms culture and change.
2 Describe and discuss the various ways of implementing change.
3 Evaluate change and the need and use of power in organizations.
4 Discuss the influence of politics in an organization.
5 Explain the importance of managing micropolitics in organizations.
6 Evaluate the cultural changes in a group that you have been a member of for twelve months or more.

CHAPTER 12

Control

CHAPTER OBJECTIVES

Define the term control
Evaluate quality control systems
Describe control process requirements
Evaluate performance indicators and their relationship to quality
Explain the need for materials control and prevention
Describe the seven old and new tools of quality

CHAPTER OUTLINE

- **12.1** What is control?
- **12.2** Quality control systems
- **12.3** Control process requirements
    - 12.3.1 *Choosing what to control — the subject*
        - Performance indicators and their relationship to quality control subjects
    - 12.3.2 *Developing a target for a control characteristic*
    - 12.3.3 *Determining a unit of measure*
    - 12.3.4 *Developing a means to measure the control characteristic*
    - 12.3.5 *Measuring the characteristic in the production arena or field*
    - 12.3.6 *Evaluating the difference between actual and expected performance*
    - 12.3.7 *Taking action as necessary*
- **12.4** Materials control and prevention
    - 12.4.1 *JIT methods*
- **12.5** The seven old and seven new tools of quality management
    - 12.5.1 *The seven old quality tools*
        - Flowcharts
        - Check sheets

Histograms
Cause and effect diagrams (fishbone diagrams)
Pareto diagrams
Scatter diagrams
Control charts
12.5.2   *The seven new quality tools*
Affinity diagrams
Interrelationship diagraphs
Tree diagrams
Matrix diagrams
Matrix data analysis
Arrow diagrams
Process decision programme chart
12.6   Chapter review
12.7   Chapter questions

12.1 What is control?

Control is the process that ensures that targets are met, through information developed in the actual process performance. It means process information being compared to expected standards and decisions made from the result. Is control necessary for good quality control? Very simply, if control were not used as the basis for all quality management decisions, then management could not manage quality at all. The concept of control is used here to provide the basis for assuring that outputs meet intended specifications and standards. The *control system* must be seen to have three components — a standard to achieve, a means of using some measure to effect an evaluation of performance and a process of comparing actual with planned results.

12.2 Quality control systems

Terry and Franklin (1982) established that there are different types of control. These include:

1   *Preliminary control* — Involves the development of measures that seek to ensure that input quality of material satisfies the required specifications; that workers know their responsibilities and can use quality control techniques; that individuals are trained effectively to carry their responsibilities; that machines and other plant and tools are available as required to ensure that product is made according to specifications; and that the production process

is designed to be as effective as possible. It also includes the managerial activities to ensure the quality of product/service produced and/or delivered.

2   *Concurrent control* — Involves the use of managers directly in the management of effective operations. It means directing the operation according to planned requirements. In a hierarchical organization, it would generally mean supervisors directing workers' tasks and therefore their outputs. In the quality-oriented organization, it usually means the workforce themselves carrying out this management task, under the guise of self-management. It also means duly changing the work environment according to immediate needs, rather than waiting.

3   *Feedback control* — Involves the use of targets and results to provide the basis for change, improvements or continuing actions. This control method is cyclical in nature. End results or outputs are used as a guide to future actions of improvement. It is a case of waiting for an output and finding out if that output satisfies the required specifications.

If it does not, then remedial action is necessary and the result must be fed back to the beginning of the process to make the required changes to that process. However, problems arise, because it must be determined whether the process is at fault or if the problem is with the raw materials or with the machinery or plant involved in the process. Either way, a delay is inevitable, which can cost money, time, effort and possibly customers.

The meaning of on-line and off-line control needs to be recognized. Production workers that use control measures on-the-spot apply on-line control measures, whereas planners use off-line control measures. On-line control tends towards the concurrent approach, where staff work to improve the present operational system, and is therefore short-term oriented. Examples include quality control charts and automatic devices. Off-line control is much longer term, and is the responsibility of management rather than the worker, as it affects the future technology uses, process and product developments and training requirements of staff. Examples include development changes in product design, and control systems that evaluate incoming materials.

Control is essentially a reactive measure. Its application seeks to ensure that process outputs to some degree match the specifications planned. Since much of quality control requires direct action at the point of production, then quality control requires production workers to exercise self-control over the process. That is, control which is on-line and instantly determining what decisions are necessary, to ensure that the product is manufactured according to derived specifications; this creates the traceable control link of performance to specifications. In this regard, workers are required to have knowledge of what they are supposed to do (their tasks and goals), knowledge about the process under their control (quality control techniques) and some means for evaluating the data generated from their process, the sort of decisions that must be made, and the authority in order to carry them out. Any quality problem can be perceived as being either management or worker controllable in terms of whether the above criteria are met.

Juran and Gryna (1993) indicate that classical control (during process execution) and self-control (preprocess execution) are closely related, but differ in terms of timing. The degree to which workers are allowed to exercise this timing over the processes they work on determines whether they can control the product they make effectively.

12.3 Control process requirements

Control involves certain requirements:

1. Choosing what to control — the subject.
2. Developing a target for a control characteristic.
3. Determining a unit of measure.
4. Developing a means or sensor to measure the control characteristic.
5. Measuring the characteristic in the production arena or field.
6. Evaluating the difference between actual and expected performance.
7. Taking action as necessary.

12.3.1 *Choosing what to control — the subject*

The actual technology of products provides a ready means for the provision of *subjects*. The major problem here is that the focus of measurement rests squarely with the product itself. Effectively, the interpretation of these derived data means that the worker is addressing the symptoms rather than the cause. The subject is therefore used as a vehicle to indicate what is happening elsewhere. The choice of the subject is therefore rather important, as it is generally used to indicate quality problems.

Quality subjects include any from the input (raw materials), process (equipment, conditions and product) and output (customers, standards authorities, etc.). Managerial issues (financial, customer and employee relations as well as performance) form the basis for the choice of subject. As quality is a customer-oriented process, so must the choice of a control subject. This is both internal and externally oriented.

An effective control subject should also:

1. Be recognized by the worker trying to generate the respective quality data as being useful and valid.
2. Be capable of being measured effectively. That is, the application of a measure produces expected results consistently.

Performance indicators and their relationship to quality control subjects

The control subject is similar to providing performance indicators (PIs). Performance indicators are defined in Hulme (Levacic, 1993) in terms of the US State Education Assessment Center. Here, the example of Blanke suggested that PIs are *inter alia*, 'a measure that conveys a general impression of the state or nature of the . . . system being examined'. PIs therefore are just that — *indicators* of performance. As Hulme further suggests 'available measures are likely to remain imperfect indicators, rather than precise guides to performance'. They are therefore rather predisposed to providing benchmarks that smack of being imprecise, singular in nature and rather narrow in application in the sense of the real meanings attached to them.

The effect of each indicator would be to provide a window on performance to determine visibly performance outcomes. It is just these factors that contribute to the almost inescapable conclusion that PIs are by themselves relatively useless, but together with others provide some understanding about the overall nature of the performance of the unit, whether in quality, financial or other non-financial terms. Consequently, it is necessary to devise, develop and use PIs that seek to cover accountable areas, and these are termed targets.

It seems relatively easy to develop PIs regarding controlled inputs such as resources (money, production material and processes) and staff. It has been easier, say, for accountants to develop and use PIs for controlled items than for non-controlled items, e.g. those measures often used in financial ratio analysis. PIs need to be deterministic, actionable, measurable and specific, just like objectives. It would seem that no one area could provide a sufficient overview of performance, and therefore it is likely that PIs would be developed throughout the range of activities of the process.

Wallace (1993) indicates that 'increasingly senior managers are realising the importance of analysing the numbers that add up to the bottom line and evaluating the activities and processes behind them'. This is seen as an important factor in this area. Drury (1992) states that 'an effective operational control and performance measurement system should emphasise both financial and non-financial measures' and predominantly quality characteristics are being evaluated on the basis of non-financial measures, and it is with these measures that many quality-related decisions will be made.

12.3.2 *Developing a target for a control characteristic*

The control subject provides the basis for developing a target for a control characteristic. When developing this characteristic, it is essential not only to accept the criteria for setting planning goals (see Chapter 5), but also to consider performance history in the organization, especially the production environment, management commitment and external factors relating to customers, etc. Juran and Gryna (1993) indicate that a number of factors need to be considered when developing a target for a control characteristic. These include:

1. All targets need to have parameters that are customer-oriented; this includes the immediate and final customers. Essentially, this means providing an *area of focus* so that greater detail can be developed later.
2. The target should be self-illuminating, in that the totality of its application will ensure that the target is representative of the measurements it will assist in providing, and that effective problem detection can be elicited from the results.

Juran and Gryna (1993) warn that 'quality control subjects (targets) must be viewed by those who will be measured as valid, appropriate, and easy to understand when translated into numbers. These are nice notions, surely. But in the real world they can be pretty elusive'.

12.3.3 *Determining a unit of measure*

The criterion of output is always some form of deficiency, because deficiency provides the basis for defect rates and possible failure problems later. However, units of measure should not just be targeted at internal requirements — a production-based quality view — they must also be used in the field. This means that different measures will be adopted, depending upon the scope of staff responsibility and their job requirements and content.

It is easier to create units of measure for tangible characteristics, such as the number of units produced, sales volumes, product units being serviced, etc. However, creating units of measure for product attributes or service is much more difficult, although service characteristics tend to be evaluated on the more tangible side. For example, customer services could use a unit of measure such as the number of clients seen in an hour, but this bears little significance to the quality of interaction between the client and customer services personnel.

What is fundamental is that no one unit of measure will suffice, and this is more true for services than for production facilities or both their consequent outputs.

12.3.4 *Developing a means to measure the control characteristic*

Measurement of a control characteristic may seem simple enough, but there are some associated problems. The job of the sensor or measuring instrument is to provide timely and effective information about the target. Consequently, being able to extrapolate outwards is vital in order to understand the situation in regard to the subject.

In the production arena, there is a growing trend towards using automatic devices to count or measure specified parameters within production runs, and any associated defects. This is also being applied to some small degree to services. However, these devices automate very simple measures. They do not of themselves provide any sophisticated means to interpret the results effectively, which is left to people. These measurement procedures account for the use of technologies such as infrared counting or measuring devices (contained in stable processes), and the use of on-line recording of conversations between customer service agent and client (subsequently analyzed for differences between the conversationally expected standard and that achieved by the service agent). Again, the basic requirement is to ensure that generated data are both accurate and timely, in order for decisions to be made on the quality of product manufactured or service delivered.

12.3.5 *Measuring the characteristic in the production arena or field*

Where do you apply the measuring sensors in the production arena or the field? Do you apply in-process (in production) or do you apply them at the end of the process, or the end of the line? The position of the sensor provides an ability to ensure that defective product does not gain any further added value.

The principle to apply here is that product needs to be measured whenever it is about to enter another part of the process. Conceivably, this will mean measuring stations at the ends of processes. But what about processes that are engineered to be very long? In these cases, interim measuring stations need to be set up. A guide could be the amount of added value given to the product as it is processed. This may be in terms of financial loss, machine time or a host of other attributes that can be effectively applied.

12.3.6 *Evaluating the difference between actual and expected performance*

Where simple measures are adopted, i.e. whole figure responses, it is relatively easy to determine whether a response is acceptable or not. Where woolly attributes are used, this may create a grey area where statistics may need to be applied in order to provide an effective means of evaluation. This statistical approach, although effective, is fraught with difficulty, as the output requires interpretation.

The difference from standard is needed and the application of a process to ensure that the developed product control measure is as effective as possible. It is the interpretation that provides the basis for decisions, not the result from the measure, and therefore the method used to effect this must be able to detect the real differences between the expected standard and that delivered. The significance of determining the differences accurately therefore cannot be overstressed. Chapter 13 on control chart development and usage may be of use here.

12.3.7 *Taking action as necessary*

Any action taken can only be carried out after ascertaining whether the highlighted problem is randomly created or can be assigned. This requires the development of problem-solving skills and the possible use of the seven old tools of quality, as discussed in section 12.5. Some results require no action at all, and this seems a reasonable evaluation, since the application of control techniques will actually result in the need for fewer reactions to the incoming data. However, in the beginning, especially when setting up the control process for the first time, the majority of results may require control decisions to be made to change the process in some objective way.

12.4 Materials control and prevention

Control of materials and inputs to processes are vital elements in the management of control of quality. Deming's fourteen points (see Chapter 3) and the application of BS EN ISO 9000 (see Chapter 15) suggest that management of incoming material should mean the reduction of supplier contacts and therefore the application of greater control measures on a lower spectrum of material sources.

12.4.1 *JIT methods*

Just-in-time (JIT) methodologies provide a basis for managing inventory and stock in order to minimize the problems of waste, as does the effective development and use of quality management standards. Vendor relationships are therefore critical to the effective management of on-line quality.

JIT means providing the right part, at the right time, in the right place. It means being able to forecast material requirements and organizing vendor shipments to ensure timely and effective supply. It means lower inventory and in some cases no inventory at all. Feedback from operations is required in such time as the vendor can supply the part as it is needed. This feedback requirement means that vendors are usually connected by computer to the customer's shop-floor computer and monitor the production line to ensure speedy and timely delivery of material. JIT puts pressures on the management of purchasing operations. It affects purchasing in the following way by making the organization:

1. Develop procedures to assure that product and process specifications are met every time.
2. Ensure that vendors are aware of their responsibilities as far as product quality is concerned and that instructions are clearly understood.
3. Ensure that vendors are not just chosen on the basis of price, but on the demonstrated consistent quality of product produced.
4. Ensure that vendors are seen as partners in the management of quality in both the operation at the vendor and at the purchaser's site.

Vendor evaluations through third party assessments based on BS EN ISO 9000 are becoming the norm. Older but nevertheless effective secondary party methods like the Ford 101 and British Gas quality systems, which imposed rigid standards of supply processes, needed to be adhered to in order to be allowed to contribute to their business. This had the effect of the customer evaluating a supplier's management system in order to assure the consistency of final product supplied. The relationship of JIT to vendor assessment is clearly demonstrated when speed and consistent quality of parts need to be supplied in a fast-moving production environment. If a part fails, then major disruption of production can occur, with its consequent delays and inherent costs. Speed of delivery of product, as and when required, must be assured. Thus, speed cannot be compromised with consistent delivery of quality.

Where the two match, the vendor is assured of survival by providing the required product and service, at the correct time and in the correct place.

It would seem that customers inject a similar cultural approach in the vendor's approach to quality. In this regard, the quality revolution will extend from the final customer to producer and on to the supplier. This chain therefore becomes synonymous with a quality link that encompasses a strategy to satisfy customers consistently down the line.

12.5 The seven old and seven new tools of quality management

The tools of quality management provide a means for individuals and groups to implement quality control processes, monitor those processes and solve any problems that derive from such processes.

The seven old and seven new tools provide an extensive armoury of weapons in order to control quality. These tools are equally applicable to both manufacturing and service-oriented organizational processes. Some of the tools are very simple in operation, but they provide invaluable data on which to make quality-related decisions. As an output from their use, the tools provide a basis for quality improvement of processes.

12.5.1 *The seven old quality tools*

The seven old tools are flowcharts, check sheets, histograms, cause and effect diagrams, Pareto diagrams, scatter diagrams and control charts.

Flowcharts

Flowcharts are relatively simple devices that illustrate the *flow* of the process being examined by showing the sequence of events for that process. They are particularly useful in understanding the input, process and output configurations.

238 THE FIVE FUNCTIONS OF QUALITY MANAGEMENT

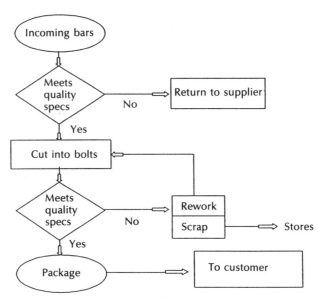

Figure 12.1 Flowchart.

Designed processes have a nasty habit of changing in the face of practical use and thus flowcharting can pin-point the change from design standards and evaluate the effect of those changes. It can also be used as a tool to develop the effective design of processes, as well as improve processes by testing different configurations prior to making a formal decision.

When flowcharting is used by the people who work with the process under examination, it provides a basis for a more objective and mutual process understanding, improved communication and a developed sense of process ownership. An example of a flowchart can be seen in Figure 12.1.

Check sheets

Check sheets are used for data collection purposes. They involve frequency determinations and sometimes tabulated dimensions are used. The process is simply to count the frequency of a given category and tally it up. The category used could be either a variable or an attribute. An example of the use of check sheets is illustrated in Table 12.1.

The result of the check can be used to develop a histogram. However, interpretation of the result, especially in complex situations, may be difficult. Have the data been collected as prescribed? Do the data have a bias, for example? Bias can mean many things, but here it encompasses the following:

1. *Interaction bias* — Where the process of collection interferes with the collected data. This can imply that the Hawthorne effect may affect the performance of the process.

CONTROL

Table 12.1 Check sheets

It is required to find out the status of the order-to-purchase delivery times from a supplier. The stated question that needs to be answered was:

How many working days does it take to deliver an ordered item?

The data set recorded is seen below:

Working days

1	2	3	4	5	6	7	8	9	10
				X					
				X	X				
				X	X				
			X	X	X	X			
			X	X	X	X			
		X	X	X	X	X			
X		X	X	X	X	X	X	X	

An X is entered when confirmation of delivery is given

2 *Procedural bias* — Where the collector does not follow prescribed means for generating the data.
3 *Mathematical bias* — Where the methods or formulae used give specific measurable bias, even within the parameters of their use.

Histograms

Histograms are a graphical representation of a given data set, and are used to visualize the data generated in check sheets, for example. Histograms are therefore useful for evaluating the pattern and shape of the distribution that reflects the population the data were drawn from. Decisions can therefore be made on the basis of this. Histograms can also be used with the designed specification limit and thus non-conforming product sample results can be clearly seen. An example of the use of histograms is illustrated in Figure 12.2.

Cause and effect diagrams (fishbone diagram)

Solving the cause, rather than the symptom of any highlighted problem, is the direct object of cause and effect diagrams. Having used the previous quality tools to develop data about the process in question and to identify the resulting problems (the effect), the cause is required in order to correct them permanently.

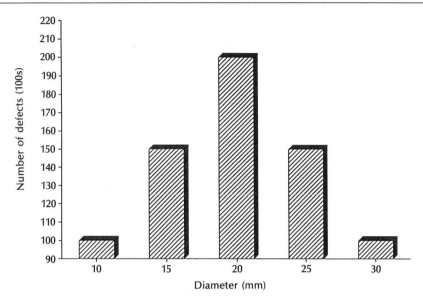

Figure 12.2 Histogram.

The diagram is essentially a set of branches (four Ms — Machines, Materials, Men and Methods) that are drawn towards a specific problem statement. Generally, more than one statement will be evaluated, thus constructing multi-fishbones. These provide a standardized platform on which to develop multi-perspectives about the causes of the various highlighted problems, normally referred to as dispersion analysis. This diagram can be equally applied to processes, where the same method of analysis applies.

Brainstorming forms the basic behavioural technique behind the analysis. An example of the use of the cause and effect diagram is illustrated in Figure 12.3.

Pareto diagrams

Pareto analysis is an attempt at reducing the focus of attention to what Juran would call the *Vital Few*. Data from check sheets could also be evaluated using this method, where the data are arranged from highest to lowest frequency. To complete the Pareto diagram, a cumulative frequency line is drawn in order to indicate the relative magnitude of the defect counts. Essentially, the Pareto configures as the 80/20 rule, where, for example, 80 per cent of defects result from 20 per cent of the available or identified causes. This usually means that a significant amount of defects can be attributed to a few evaluated categories and it is in these categories that attention should be focused — which would result in a significant quality improvement in the process. An example of this can be seen in Figure 12.4.

CONTROL 241

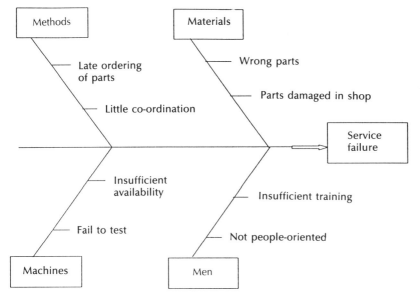

Figure 12.3 Fishbone diagram.

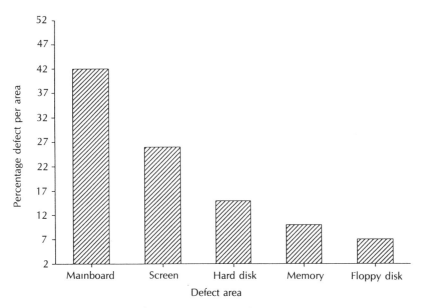

Figure 12.4 Pareto diagram.

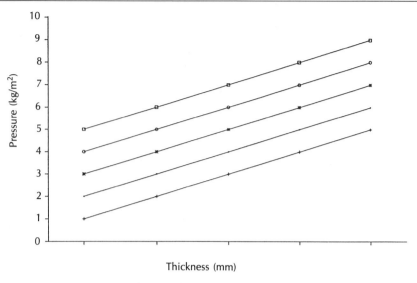

Figure 12.5 Scatter diagram.

Scatter diagrams

Scatter diagrams are based on the application of regression analysis and the result displayed in graphical form. The diagram is set up by evaluating the relationship between two sets of variables. The trend or statistical correlation developed by the regression analysis provides a basis for the interpretation of the diagram. Relationships are determined on the basis of whether there is positive ($+1$), negative (-1) or no correlation (0). An example of the use of a scatter diagram is illustrated in Figure 12.5.

Control charts

A control chart is a graphical display of a quality characteristic that has been measured. These are discussed in much greater detail in Chapter 13.

12.5.2 *The seven new quality tools*

These tools are much more complicated than the previous old tools of quality. They are more functional, in terms of engineering principles and practices and therefore their popularity has not yet been confirmed by business managers.

Affinity diagrams

The affinity diagram is used to generate a large number of ideas and facts relating to a stated problem area. This method is based on the development of related patterns and groupings.

Interrelationship diagraphs

This technique provides a means of taking a fundamental idea and developing logical links from among seemingly related categories by the use of lateral thinking processes. This technique is generally used after the affinity diagram.

Tree diagrams

Tree diagrams map out the paths and relevant associated tasks in order to accomplish a given project.

Matrix diagrams

Matrix diagrams develop graphical relationships between characteristics, functions and tasks and they do so sequentially by providing logical connections. The House of Quality is the best example of the use of this technique.

Matrix data analysis

This technique takes relevant data from the matrix diagram and graphically displays their corresponding relationships in terms of quantity and strength. It is essentially a factor analysis technique.

Arrow diagrams

These are PERT and CPM planning techniques. Their use has been generally confined to construction or heavy engineering industries.

Process decision programme chart

This technique is used to quantify each possible and probable event and their solutions. It is a proactive measure that seeks to determine effective countermeasures to anticipated problems.

12.6 Chapter review

Control is the process that ensures that targets are met, through information gathered in the actual process performance. It means that the process information is compared to what was planned and decisions are made on the result.

Different types of control system are preliminary control, measures that seek to ensure that input quality of material satisfies required specifications; concurrent control, which involves managers directly in the management of a given process operation; and feedback control, which involves the use of targets and results to provide the basis for change, improvements and continuing actions.

Control involves choosing what to control, developing measurement characteristics and techniques, measuring, evaluating and taking the necessary action as a result.

12.7 Chapter questions

1. Define the term control.
2. Evaluate quality control systems.
3. Describe control process requirements.
4. Evaluate performance indicators and their relationship to quality.
5. Explain the need for materials control and prevention.

CHAPTER 13

Statistical process control

CHAPTER OBJECTIVES

Define the term statistical process control (SPC)
Evaluate the process of inspection
Discuss the concept and development of statistical process charts
Discuss quality control charts
Evaluate the basis and use of acceptance sampling techniques
Evaluate the impact of acceptance plans
Describe single and double sampling plans
Evaluate the operating characteristic curve
Describe process capability and its effective use
Evaluate the cumulative summation (CuSum) control chart
Compare control and CuSum charts
Describe the advantages and disadvantages of control charts
Describe the advantages and disadvantages of CuSum charts
Evaluate the relationship between SPC and quality improvement
Evaluate process variability reduction

CHAPTER OUTLINE

- **13.1** Introduction
- **13.2** What is statistical process control (SPC)?
- **13.3** The process of inspection
    - 13.3.1 *Measurement*
    - 13.3.2 *Quality measurements*
    - 13.3.3 *Variation*
- **13.4** Statistical control charts
    - 13.4.1 *The concept of the control chart (Shewhart control chart)*
        Definitions
        Application of the control chart
    - 13.4.2 *Steps in the development of a control chart*
- **13.5** Quality control charts

        13.5.1 *Variable control charts*
        13.5.2 *Attribute control charts*
13.6   Cumulative summation control chart (CuSum)
13.7   Comparison of the control and CuSum charts
13.8   Advantages and disadvantages of control charts
13.9   Advantages and disadvantages of CuSum charts
13.10 Acceptance sampling
13.11 Acceptance plans
13.12 Sampling plans
        13.12.1 *The single sampling plan*
        13.12.2 *The double sampling plan*
13.13 The operating characteristic curve
        13.13.1 *Constructing the operating characteristic curve*
13.14 Process capability
        13.14.1 *Process capability uses*
13.15 SPC and quality improvement — what it means
13.16 Process variability reduction
13.17 Taguchi — the quality loss function
13.18 Chapter review
13.19 Chapter questions

13.1 Introduction

Statistical process control (SPC) provides the means for managers, through worker operatives, to administer and manage their organizational processes in a way that ensures that goods and services satisfy customer needs and wants effectively.

    SPC means the adoption of techniques based on an effective mathematical appraisal of a given process in order to assure the product is produced as designed, with minimum wastage.

13.2 What is statistical process control (SPC)?

SPC is the application of statistical techniques to a process in order to:

1   Develop and collect statistical data about a process.
2   Apply these techniques to provide the basis for interpretation of the functionality and performance of that process.

SPC involves the measurement of data surrounding variation in a process and can be applied to any part of that process. In this sense, a process involves a *combination* of materials, technology, resources and methods that bear on an organization's ability to produce goods and services that satisfy those stated or implied customer needs and wants.

The Japanese Industrial Standards define quality control as a 'System of production methods which economically produces quality goods or services meeting the requirements of consumers' (Ishikawa, 1985).

Here, we will consider the use of SPC in managing variation and the development of the provision of statistical techniques that allow individuals working on the process to manage its output effectively. It has to be said that without the consistent application of SPC to a process, the quality of a product or service cannot be guaranteed and the quality objectives may not be continuously achieved.

However, statistical techniques do not in themselves provide the quality of product demanded by customers, but do provide a methodical means of assuring that designed specifications are consistently met. These techniques demand education and training, not only of the managers but also — and possibly as important — of the individuals on the production line.

13.3 The process of inspection

13.3.1 *Measurement*

Statistical techniques rely on one major requirement — data. What do you actually measure to generate the data in order to apply statistical techniques? Measuring focus includes any element or defined quality feature that can be evaluated and/or appraised.

Measures are usually adopted in terms of functional requirements, measures that purport to relate directly to the output performance characteristics of the process, e.g. measures that reflect whether dimensions or specifications of the product/service are within planned set tolerances. This is because variation and the application of statistical techniques are about evaluating the exception, rather than the norm. Consequently, measures generally reflect deficiencies in the process performance.

Measures will therefore usually be directed to ensuring that the process produces a product consistently to specifications. This means that each item produced will be evaluated according to the specifications set. However, economics normally prevents the evaluation of each product. This is where statistical quality control will help for both requirements. SPC will ensure that a sample of product representative of all product produced can be selected and evaluated. SPC can also provide the basis for evaluating whether the process itself is statistically within control. Therefore, measurements using developed data provide the basis for ensuring that the process manufactures products and/or services to specification every time.

13.3.2 *Quality measurements*

Examples of quality measurements include:

- MANUFACTURING

Production
Defect rates per thousand items produced.
Defect rates per sample.
Defect rate of scrap and rework (in terms of cost or number of items).
Defect rate of product getting to the customer.
Production downtime.
Input materials/product defect rate.
Percentage of specification design changes.
- SERVICES
- SERVICES

Internal
Percentage production downtime.
Defect rate of procedural errors.
Percentage rate of orders processed.

External
Product/service delivery time.
Order response time.

Much of the measurements used here are reflections of work outputs — whether in-process or at the end of the process.

Consequently, a major problem of measurement implementation is from the inherent resistance of workers and unions to using personal performance measures as part of the process performance. This generally requires effective negotiations in order to implement the measures successfully. Many managers mistakenly believe that one measurement is sufficient to evaluate a process and the products that derive from it. This means that more than one measure should be used and a choice must be made between competing measures as to their worth and effectiveness.

13.3.3 *Variation*

The understanding of variation is an important concept to grasp when managing process quality. Variation in practice results in non-identical production of a product or service. It is the nature of variation that means that some form of applied management of the process must be accomplished in order to minimize this inherent variation. Statistics provides the basis for this management and thus helps managers make the correct decisions about product service quality.

13.4 Statistical control charts

Modern quality control or statistical quality control began in the 1930s with the industrial application of the control chart, which seems to have been attributed to Dr W A Shewhart (1931) of the Bell Laboratories. The Second World War was the catalyst that made the control chart's application possible to various industries in the United States and Europe by default rather than by the application of managerial evolutionary developments. Modern quality control practices use statistical methods that assist in continuous improvement drives.

To ensure customer requirements are met after design, production methods require some form of control. Increasingly in manufacturing — especially mass production — and in service industries, SPC methods are becoming the norm rather than the exception. Indeed, to minimize costly waste, a control method is essential. One method that allows manufacturers and service organizations to control their processes is the simple control chart.

13.4.1 *The concept of the control chart (Shewhart control chart)*

There seems to be no such thing as constancy in real life. There is, however, such a thing as a *constant cause* system (Deming, 1986). The results produced by a constant cause system vary, and in fact may vary over a wide or narrow band, but they exhibit an important feature — stability. Why apply the terms constant and stability to a cause system that produces results that vary? Because the same percentage of these varying results continues to fall between any given pair of limits so long as the constant cause system operates. It is the distribution of results that is constant or stable.

Definitions

> When a manufacturing process behaves like a constant cause system, producing inspection results that exhibit stability, it is said to be in statistical control. The control chart will tell you whether your process is in statistical control or not. (Deming, 1986)

> A control chart is a graphic comparison of process performance data to computed "control limits" drawn as limit lines on the chart. (Juran and Gryna, 1993)

> A control chart is a graphical display of a quality characteristic that has been measured or computed from a sample versus the sample number or time. (Montgomery, 1985)

Application of the control chart

The chart detects the presence of sources of variation due to specific assignable causes. These assignable causes of variation are in contrast to random causes, i.e. those solely due to chance. If we can detect and eliminate assignable causes, it can help to prevent defects occurring. If data are collected which have a bearing on any problem, any series of events or any manufacturing or service situation, these data are **always** found to exhibit variation. Instead of being exactly the same from point to point or from time to time, the numbers vary. If plotted on graph paper so that the variations can be studied, the numbers always form a fluctuating zig-zag pattern. This applies to any process under consideration, including accounting figures, production figures, records of attendance, temperature, etc. Occasionally there will be a large or unusual difference, much more important than all the other causes combined. These large causes make the pattern fluctuate in an unnatural way.

Montgomery (1985) suggests that control charts:

1. *Are a proven technique for improving productivity.* This is done by optimizing the process by continuous surveillance of its output and determining its conformance to set production limits.
2. *Are effective in defect prevention.* As stated above, this surveillance ensures that special causes can be detected before product is produced outside set production limits.
3. *Prevent unnecessary process adjustments.* Whilst in theory this may be so, in practice the operator may well have to be trained to prevent him/her from adjusting the process unnecessarily.
4. *Provide diagnostic information.* The trends and direction of the points plotted will illustrate the variation pattern of the production process.
5. *Provide information about process capability.* The continuous analysis of process capability will prove that product is produced consistently, or will assist in the development of new process and product standards.

Control charts are used to provide a visualization of the data generated from evaluating the process and products from that process. Control charts are therefore graphical in nature, and represent the performance characteristics measured. They are therefore limited in helping to make decisions by the very nature and content of the measurements that are made.

The chart is designed using design specifications (tolerances) or data groups (collected groups of samples) that are assigned positions on the chart after computation. An example of a chart is contained in Figure 13.1. The chart is used to provide a means to detect *special* or assignable causes of variation. Past data are computed to provide demonstrated limits of the process and help develop a predictable process.

The chart illustrates the relationship between three lines. These are:

1. *Upper control limit* — the level above which it is assumed that the process is out of control.

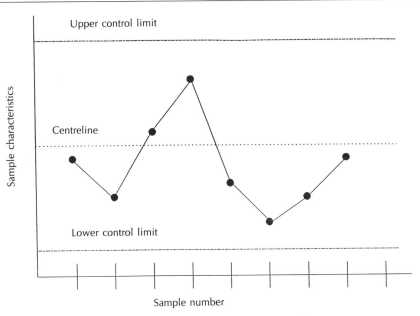

Figure 13.1 Typical control chart.

2   *Lower control limit* — the level below which it is assumed that the process is out of control.
3   *Centreline* — average of all the individual samples (at set up).

Variations can be given to two kinds of causes — special or common. Common causes of variation happen at random and there is little that can be done to alleviate them. However, special or assignable causes are attributable to causes that can be minimized or eliminated. In the best managed processes, only common causes are present. Control charts are therefore used to manage special variation effectively.

It is the computation of control limits that distinguishes between common and special causes of variation. The skill lies in making or ensuring the control limits are wide enough to allow all special causes to be ascertained and eliminated, whilst reducing the effect of common causes. The control chart detects the changes in variation, but does not point to the actual cause.

This is a limiting feature of control charts, but one in which the experience of managers and operators, with intimate knowledge of their processes, can actually reduce and overcome. Once the control chart has been set up, it is considered that only special causes affect the process under consideration. If a data result is seen to exceed the control limits, then it is assumed to be as a result of special causes, and therefore the cause must be sought and eliminated.

13.4.2 Steps in the development of a control chart

Step 1 — *Choose the measurement feature or performance characteristic.* Specific attention should be given to characteristics that are seen to be problematic, important or critical. This must be done in a way that ensures that the individual operator can actually measure and complete the data generation task as part of the normal operation. Process and product/service output variables need to be carefully chosen as the input and process variables directly contribute to output performances.

Step 2 — *Choose measurement methodologies* that will not only allow the effective management of the process to be achieved, but also the ability to pin-point problems and diagnose their causes.

Step 3 — *Choose the type of control chart to be used*. Variable (X bar and R chart) or attribute (percentage or number non-conforming chart).

Step 4 — *Decide the basis for a centreline and the calculation of the control limits*. Centrelines are best developed using past data. However, when implementing a new process or product, then designed centrelines and control limits will be used (called tolerance limits).

Step 5 — *Decide on the sample size or sample grouping*. Sample size reflects the product throughput, as well as the speed of this throughput and the time available for calculating and registering the sample result. It is necessary to ensure that differences between individual samples in the same subgroup are minimized, and that the differences between subgroup results are maximized.

Step 6 — *Design a system to collect data*, which must be as simple as possible. This includes designing the use of instruments that ensure an error-free indication and recording of the status of the product characteristic under measurement. The use of computers should be seen as a means to ensure reliability of data generation as well as providing an equitable basis for data analysis. The simplest system will be integrated by using the same technology to generate, record and analyze data, and output these in a convenient form for the worker. Cognisance must be taken of the environment in which the worker operates — especially in relation to the transfer of data from the measurement stage to the analysis stage.

Step 7 — *Calculate the control limits*. Formulae used in the development of the control limit can be seen in Table 13.1.

Step 8 — *Plot and interpret results*.

The advantages of using charts are that they:

1   Can be used as process improvement tools.

Table 13.1 Formulae table

Chart type	Centreline	Lower limit	Upper limit
Average	$\bar{\bar{X}}$	$\bar{\bar{X}} - A_2\bar{R}$	$\bar{\bar{X}} + A_2\bar{R}$
Ranges	$\bar{R}$	$D_3\bar{R}$	$D_4\bar{R}$
Percentage non-conforming	$\bar{p}$	$\bar{p} - 3\sqrt{\bar{p}\left(\dfrac{1-\bar{p}}{\bar{n}}\right)}$	$\bar{p} + 3\sqrt{\bar{p}\left(\dfrac{1-\bar{p}}{\bar{n}}\right)}$
Number of non-conformities	$\bar{c}$	$\bar{c} - 3\sqrt{\bar{c}}$	$\bar{c} + 3\sqrt{\bar{c}}$

2   Provide a ready means for process problem detection and diagnostics.
3   Provide data about the process capability.

13.5 Quality control charts

Quality control charts are divided into two main types, depending upon the characteristics being evaluated: variable control charts and attribute control charts. For the benefit of our discussion $\bar{X}$ and R charts are considered in variable control charts and $\bar{p}$ and c charts for attribute charts.

13.5.1 *Variable control charts*

In the variable chart, a small subgroup sample is taken from the stream of product produced at predetermined and regular times. The average and the ranges of the subgroup sample are calculated. A minimum number of samples — at least 30 — are taken from the production stream to provide the basic data about the process. This means that a maximum of 150 actual product samples are taken for evaluation (30 × up to 5 per subgroup).

The control limits are set using $\pm 3\sigma$ for the $\bar{X}$ chart, which is plotted on a separate chart to the R values.

The formulae for both charts are:

	Centreline	Lower limit	Upper limit
Average	$\bar{\bar{X}}$	$\bar{\bar{X}} - A_2\bar{R}$	$\bar{\bar{X}} + A_2\bar{R}$
Ranges	$\bar{R}$	$D_3\bar{R}$	$D_4\bar{R}$

Where, for the calculation of control limits:

$\bar{\bar{X}}$ = average of the sample averages
$\bar{R}$ = average of the sample ranges
A_2, D_3, D_4 = constant based on n, number of subsamples (see Table 13.2).

Table 13.2 Constants used in constructing X bar and R charts

n	A_2	D_3	D_4	d_2
2	1.880	0	3.268	1.128
3	1.023	0	2.574	1.693
4	0.729	0	2.282	2.059
5	0.577	0	2.114	2.326
6	0.483	0	2.004	2.534
7	0.419	0.076	1.924	2.704
8	0.373	0.136	1.864	2.847
9	0.337	0.184	1.816	2.970
10	0.308	0.223	1.777	3.078

An example of the use of a variable control chart may help to synthesize the theory. Consider a manufacturer of pens. A major requirement is to control the inside diameter of the point on the barrel that the pen nib fits into. Fifteen sample groups (each of four samples each) were taken and it is wished to determine whether the process is in control using the $\bar{X}$ and R charts. The results can be seen in Table 13.3.

$$\bar{R} = R/10 = 1.32/10 = 0.13$$

For sample of $n = 5$, from Table 13.2, $D_3 = 0$ and $D_4 = 2.114$. The control limits to construct are:

Lower control limit $= \bar{R} D_3 = 0.13 \times 0 = 0$
Upper control limit $= \bar{R} D_4 = 0.13 \times 2.114 = 0.275$

The R chart can be seen in Figure 13.2. The R chart result indicates that the process variability is in control and we can now construct the $\bar{X}$ chart.

Table 13.3 Constructing X bar and R charts — example

Sample number	Measurements observed					$\bar{X}$	R
1	4.11	4.12	4.07	4.01	4.09	4.08	0.11
2	4.05	4.14	4.00	4.10	4.06	4.07	0.14
3	3.99	4.17	4.12	4.14	4.09	4.10	0.18
4	4.08	4.11	4.14	4.10	4.03	4.09	0.11
5	4.12	4.05	4.07	4.11	4.10	4.09	0.07
6	4.14	4.08	4.02	4.19	4.09	4.10	0.17
7	4.13	4.16	4.06	4.03	4.11	4.10	0.13
8	4.02	4.10	4.09	4.07	4.14	4.08	0.12
9	3.98	4.17	4.13	4.08	4.10	4.09	0.19
10	4.03	4.07	4.10	4.02	4.12	4.07	0.10
						Σ 40.87	1.32
						$\bar{X}$ 4.09	$\bar{R}$ 0.13

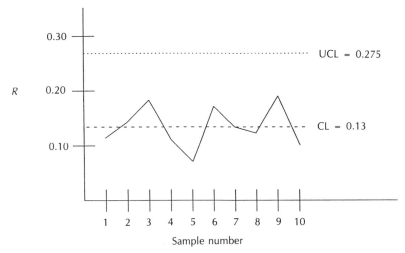

Figure 13.2 R chart — example.

For sample of $n = 5$, from Table 13.2, $A_2 = 0.577$.
The $\bar{X}$ centreline (CL) is determined as $\Sigma \bar{X}/10 = 4.09$. The control limits are constructed using the formulae:

Upper control limit (UCL) $\bar{X} + A_2\bar{R}$ $4.09 + (0.577)(0.13) = 4.16$
Lower control limit (LCL) $\bar{X} - A_2\bar{R}$ $4.09 - (0.577)(0.13) = 4.01$

The $\bar{X}$ chart can be seen in Figure 13.3.

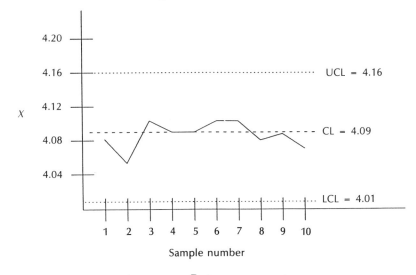

Figure 13.3 $\bar{X}$ chart — example.

Analysis of the chart reveals that no out of control points are detected and that the process is therefore operating in a state of control.

13.5.2 *Attribute control charts*

Attributes are considered to be less *objective* than variable control charts. These characteristics provide integer counts because fractions are not possible, i.e. the number of non-conforming items in a sample can be 1 or 2, but not 1.5. The essential element is that some process characteristics are not entirely quantitative in the sense of actual measurements. Consequently, counting defects in relation to a defect may be all that can be done. But, determining whether a characteristic is a defect or not is left to the discretion of the operator and is based on education, experience and skill. It is therefore reasonably subjective in nature and application. However, the calculations are just as objective as the use of the variable control chart. The calculation of the attribute control chart is very similar to the calculation and construction of the variable chart.

Attribute charts are developed to solve two independent problems and are therefore applied in two forms. The first of these is the $\bar{p}$ chart. The $\bar{p}$ chart represents the fraction non-conforming. It is calculated using the following formula (which is based on the binomial distribution):

$$\text{Centreline} = \bar{p}$$

$$\text{LCL and UCL} = \bar{p} \pm 3\sqrt{\frac{p(1-p)}{n}}$$

The sample fraction non-conforming can be determined from the following formula:

$$\bar{p} = D/n$$

where D is the number of non-conforming units in a sample size n.

In general, samples of units n are taken from a process under inspection, determining the fraction non-conforming, and plotting the result on the p chart, using the formula above. Similar principles of operation exist with this as with the variables charts; as long as the fraction non-conforming results keep within the boundaries set for the chart operation, then the process is considered in control.

An example of the use of the p chart may help to synthesize the theory.

It is considered necessary to ensure that a process is producing a product within specifications. The product is the plastic container for the ink used in a fountain pen. Inspection is necessary in order to determine whether the bottom seal will leak when filled. A control chart needs to be set up. To do this, a sample size of 24 is selected of $n = 30$, which is collected every 20 minutes over an 8 hour production run. The resulting data can be seen in Table 13.4.

Table 13.4 p — sample fraction non-conforming — example

Sample number	Number of non-conforming units	Sample fraction non-conforming p
1	5	0.17
2	8	0.27
3	3	0.10
4	9	0.30
5	11	0.37
6	6	0.20
7	3	0.10
8	9	0.30
9	12	0.40
10	7	0.23
11	3	0.10
12	8	0.27
13	2	0.07
14	6	0.20
15	10	0.33
16	5	0.17
17	7	0.23
18	2	0.07
19	9	0.30
20	11	0.37
21	13	0.43
22	8	0.27
23	5	0.17
24	9	0.30
Sum =	171	$\bar{p} = 0.2375$

The control limits can now be provisionally set as:

$$\bar{p} = \frac{171}{(24)(30)} = 0.2375$$

Using this as an estimate of the true process non-conforming, the upper and lower limits of the control chart can be developed, as:

LCL and UCL $= \bar{p} \pm 3\sqrt{\frac{p(1-p)}{n}} = 0.2375 \pm 3\sqrt{\frac{0.2375(1-0.2375)}{30}} = 0.2375 \pm 0.233$

Therefore:

UCL = 0.470 and LCL = 0.004

The p control chart is shown in Figure 13.4. Analysis of the chart reveals that no out of control points are detected and that the process is therefore operating in a state of control.

The second of the attribute charts is the c chart. The calculation of this attribute control chart is very similar to the calculation and construction of the p chart.

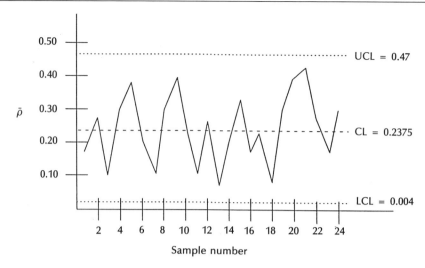

Figure 13.4 $\bar{p}$ chart – example.

A non-conforming product is any product that fails to meet any one of the specified parameters of inspection. This results in a defect. Defect analysis and detection do not mean that the product is non-conforming, as there could be a degree of latitude applied to defects that are deemed to be minor. However, too many detected defects or defects affecting critical parameters can be determined as evidence of a non-conformance. Charts can be developed for the actual number of non-conformities, or the average number of units detected in a unit and based on Poisson's distribution as there is assumed to be a constant sample size — which could be in groups of four, or five, etc. Using Poisson's distribution, the following formulae can be developed for use on the c chart:

Centreline $= \bar{c}$
UCL and LCL $= \bar{c} \pm 3\sqrt{c}$

If a negative value for LCL results, then the value 0 is used, c being both the mean and variance of the Poisson distribution.

An example of the use of the c chart may help to synthesize the theory. A manufacturer produces plastic-coated worktops. An inspection revealed the following non-conformities in twenty samples. These are contained in Table 13.5.

$\bar{c} = \dfrac{83}{(20)} = 4.15$

UCL $= \bar{c} \pm 3\sqrt{c} = 4.15 + 3\sqrt{4.15} = 10.26$
LCL $= \bar{c} \pm 3\sqrt{c} = 4.15 - 3\sqrt{4.15} = -1.96$

The c control chart is shown in Figure 13.5. Analysis of the chart reveals that

Table 13.5 c — non-conforming units — example

Sample number	Number of non-conforming units
1	5
2	3
3	3
4	6
5	7
6	1
7	3
8	5
9	4
10	8
11	3
12	8
13	2
14	6
15	2
16	5
17	3
18	2
19	4
20	3
Sum Σ =	83

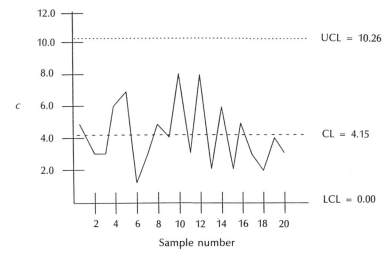

Figure 13.5 $\bar{c}$ chart — example.

no out of control points are detected and that the process is therefore operating in a state of control.

13.6 Cumulative summation control chart (CuSum)

CuSum charts were first proposed by Page in 1950 (Montgomery, 1985). The CuSum chart is a useful statistical method in analyzing data. Whenever a process is in control and the specification is written around the process average, using the CuSum statistical technique causes the data to revolve about the average value. Comparing plotted CuSum data with the specification requirement and the data averages shows that most specifications do not reflect what the process is actually doing.

To make the CuSum data plot more understandable results, analysts should be aware of two specific issues:

1   When choosing a statistical method, the measurement parameter must be critical to the manufacturing process.
2   Knowledge of the process being evaluated helps, as it allows analysts to evaluate the data with respect to the process.

Unlike the limits of a Shewhart control chart, which are fixed and parallel, CuSum charts have limits that vary in position and can be applied by laying a suitably shaped template over the CuSum chart, with a fixed origin coinciding with the last plotted point.

CuSum charts are more general than Shewhart charts, in that control can be achieved by individual readings; the sample size is $n = 1$. CuSum charts are used to detect changes or shifts of between 0.5 and 2.0 $\sigma \times$ sample mean in process level. CuSum charts have plotted points that contain information about all the observations up to and including it, i.e. the CuSum chart directly incorporates all the information in the sequence of sample values by plotting cumulative sums of the sample values from a target value, whereas Shewhart control charts contain information only from a single subgroup. The interpretation of this chart is determined by comparing plotted points with defined limits. On a CuSum chart, the mean process level is determined by the slope of the plotted points. If the process is operating with a mean equal to the target value, the slope is zero.

An example will clarify the construction of a CuSum chart. Suppose we are preparing a liquid washing soap, in which we wish the average content of a particular ingredient to be as near as possible to 20 g/l. The liquid produced will be mixed in bulk prior to the next operation, as the average must be kept at 20 g. Normally the reference value of 20 g/l is used as the control for the process, historical data help to determine this value. An example can be seen in Table 13.6.

Table 13.6 CuSum — example

No.	Observation	Reference value	Difference	Cumulative difference
1	20	20	0	0
2	19	20	−1	−1
3	21	20	+1	0
4	22	20	+2	+2
5	20	20	0	+2
6	19	20	−1	+1
7	20	20	0	+1
8	20	20	0	+1
9	18	20	−2	−1
10	21	20	+1	0
Average	20			

The process is tested at various intervals, and is systematically the reference value from each observation. What remains is then plotted as a difference up or down from the last point (see Figure 13.6). This would comprise one set of results that could be plotted on a graph, as illustrated in Figure 13.6.

Another set of results could be developed and plotted on the graph, giving a comparative indication of process performance. Obviously the curve reflects the data developed through the application of the cumulative difference. The result does indicate that there is no relative change in the process throughout the sample duration, as there is a horizontal regression line. Interpretation then requires analysis of the following:

1   The slope of the regression line. Where this line is horizontal, it means that the average value is equal to the reference value. Where the line is sloping

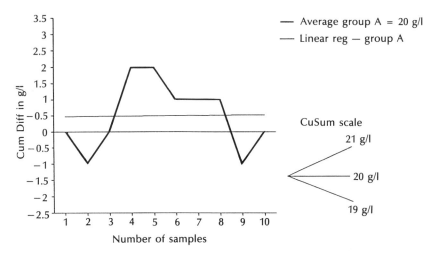

Figure 13.6 CuSum chart — example.

up or down (as in this case), a new process average has developed. The greater the difference, the steeper the line, and the greater the difference between the new average and the reference value. The change of mean in each series becomes a change of slope in the CuSum (Montgomery, 1985).

2   If the curve continues upwards or downwards through each subgroup, then the process averages are also increasing.
3   Each new point represents not only the value of the difference at that point, but also the past history of the process.

In the course of the interpretation of this exercise, there would seem to be a need to adjust the process to ensure that the process average moves in a more horizontal direction, i.e. conforms to expected process average.

A problem with using CuSum charts is determining at what angle of upward or downward slope of a process means that the process is out of control. To assist in this task, it is normal practice to derive a 'V mask' — as developed by Barnhard (1959) (cited in Montgomery, 1985) — although there are other methods developed to do similar things. However, Murdoch uses the *decision interval* control scheme, where the CuSum decision taking is based on setting the ARL at:

1   An acceptable quality level.
2   The reject level.

Montgomery (1985) discusses the use of the mask and suggests that:

> the decision procedure consists of placing the V mask on the cumulative-sum control chart with the point 0 on the last value of the sample and the line OP parallel to the horizontal axis. If all the previous cumulative sums lie within the two arms of the mask, the process is in control. However, if a sample lies outside the arms of the mask, the process is considered out of control.

Each new result would be plotted on the CuSum chart and the mask applied to determine if the process is in control. Thus, the mask forms a visual means of determining control limits that can be applied to a CuSum chart, much the same as the control limits on the Shewhart chart.

The performance characteristics of the CuSum chart are therefore dependent upon the effective design of the V mask, which can be illustrated using the formula:

We need d, the lead distance, $O = -2 \dfrac{\ln \alpha}{\delta^2}$

O, the angle of the V $= \tan^{-1}(\Delta/2A)$

Having found both the above, there is a need to calculate both H and K:

$H = 2d\delta_x \tan(\theta) \qquad K = \Delta/2$

13.7 Comparison of the control and CuSum charts

A simple way of comparing both these charts is to apply the concept of average run length (ARL). The ARL is the average number of sample points plotted, at a specified quality level, before the chart detects a shift from a previous level. Comparisons are made between the ARL resulting from both the Shewhart and CuSum charts, for various amounts of shift in a process average. For shifts between about 0.5 and $2\sigma\ \bar{X}$, the CuSum chart detects shifts with about half the number of subgroups as the Shewhart chart. The Shewhart chart would seem to be best for larger shifts, e.g. $> 2\sigma\ \bar{X}$. Both charts give a visual representation of changes in operating variables, but to different degrees.

13.8 Advantages and disadvantages of control charts

Advantages

1. Relative simplicity. Simple mathematics is involved in the construction and operation of the chart.
2. A graphical image is developed that seeks to reinforce process changes.
3. Trends of plotted data can be seen and acted upon.
4. The graph represents process capability.

Disadvantages

1. Operators need to be trained **not** to adjust the process continuously.
2. Set limits determine capability constraints on the process.
3. Plotted points indicate a shift in the process after twice as many subgroups as the CuSum chart.

13.9 Advantages and disadvantages of CuSum charts

Advantages

1. Because the chart combines previous plotted values, they are more effective than Shewhart charts for detecting small process shifts.
2. Plot points covered by the top of the mask signify a decrease in the average.
3. Plot points covered by the bottom of the mask signify an increase in the average.

4   If all plot points are uncovered, then the process is determined as being in control.
5   The CuSum chart detects movement of the process towards out of control conditions faster than the Shewhart chart.
6   CuSum charts require approximately half of the number of subgroups to detect shifts in the process average, for shifts >0.5 to $2.0\,\sigma \times$ sample mean.

Disadvantages

1   Much more complicated to determine the CuSum chart than the Shewhart chart.
2   Range chart check is required to see if the standard deviation is constant before drawing conclusions on average.
3   Gradual changes, say over five to six subgroups, are not as apparent on the CuSum chart as on the Shewhart chart.
4   More difficult to develop the mask.
5   V mask may be difficult to determine, which does not seem to be a reasonable task for the everyday worker.

13.10 Acceptance sampling

Sampling is the process of evaluating a portion of a product grouping in order to make a decision about whether to accept or reject the whole assignment. Sampling is required for various reasons and these include when the:

1   Cost of sampling the whole lot is considered too expensive.
2   Time taken to inspect all product is too long.
3   Inspection process itself requires the destruction of the product under inspection.
4   Cost of inspection is considered too high in relation to the perceived cost of passing a defective product.
5   Inspection process itself is too complicated to be carried out for each product produced.

Acceptance sampling is deemed unnecessary when the supplier produces product in a process that conforms to a quality management standard, such as BS EN ISO 9000. Acceptance sampling bears some risks that include greater administrative costs and less comprehensive awareness of total product information than would be gathered by 100 per cent inspection. However, acceptance sampling provides greater assurances when supported by a quality improvement programme that includes the use of statistical control charts.

The major requirement for acceptance sampling is essentially economic. Figure 13.7 indicates the various sampling alternatives:

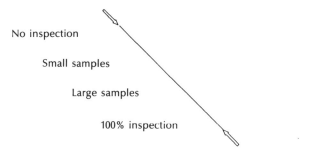

Figure 13.7 Inspection economics.

- *No inspection* — Applicable when a supplier has a quality management system in place in which delivered product is produced within this process, whether internal or external.
- *Small samples* — Ordinarily used when the process successively produces a uniform product over many items. This is the circumstance for the use of process control techniques.
- *Large samples* — Delivered product that has been produced through a process, the quality of which has not been determined, which would necessarily involve the use of this alternative. It is expensive in both the inspection and in relation to product that is defective but not detected.
- *100 per cent inspection* — This follows on from one of the last two — small and large sample inspections — if they indicate that there is a problem then a usual option is to opt for 100 per cent inspection. However, this is only a real option when the product cannot be shipped back or that the time delay between sending the product back and getting new product in return does not affect your own production schedules.

In an organization that is quality-oriented, 100 per cent or large sample inspections are not deemed realistic alternatives. No inspection is an option only if the supplier has a proven documented quality management system. Small samples are only used when the product needs to be checked for, say, health or very rigorous safety requirements.

13.11 Acceptance plans

These are sample plans that are based on arbitrary measures such as 5 per cent of lot size, e.g. if a lot contained no defects upon inspection, then the lot was accepted; if the lot contained defects, the lot was rejected. Sampling plans should be developed according to the need to reduce both the probability of accepting a defect, as well as the probability of rejecting good product.

Features of good sampling plans include:

1. The basis of the sampling plan should reflect producer and customer needs and requirements and not just arbitrary or statistical measures.
2. The operating curve should provide the basis for evaluating the development of an effective plan.
3. The plan should consider:
   (a) The cost of 100 per cent inspection.
   (b) Any data about the process, such as process capability.
   (c) Flexibility requirements of the sampling plan.
   (d) Training requirements of personnel.
   (e) The cost of implementing the plan.
   (f) The testing of the sampling plan at regular intervals, by applying known data.

It is essential that sampling plans are tested properly before use. This means piloting prospective plans against known lot and sample product data. Many characteristics affect the performance of sampling plans. Two major characteristics are the lot and sample size. Lot size is deemed to have little effect on acceptance probability, but cannot be totally ruled out when considering its effect on the sampling plan. Higher lot sizes will mean a more robust and effective operating curve. If the sample size and the acceptance number are increased in unison, the resultant operating curve (OC) becomes near ideal.

Various indices help the quality manager develop sampling programmes. These include:

- AOQL — *Average outgoing quality limit* — It is deemed that there is some sort of relationship between fraction defective prior to inspection and that after inspection. The AOQL provides a measure between two points on the quality acceptance continuum. On the one side, incoming product has no defects and outgoing product will also have no defects.

  On the other side, incoming product is high in defects, and outgoing product contains no defects. Therefore in both incoming quality cases, outgoing quality will always be good. Midway between these two points, the defective rate in the outgoing product will reach its maximum and this is the AOQL.
- AQL — *Acceptable quality level* — This is deemed to be related to the designed quality characteristics of the product or service that are offered for measurement. It is therefore particular to those characteristics of that product or service. Since the AQL is an *acceptable* level, it is notional and arbitrary, and therefore should be highly attainable. AQL is essentially defined in terms of the maximum acceptable fraction non-conforming found in a sample.

Other indices that can be used include the limiting quality level (LQL); this refers to the unacceptable level of quality so the probability of acceptance must

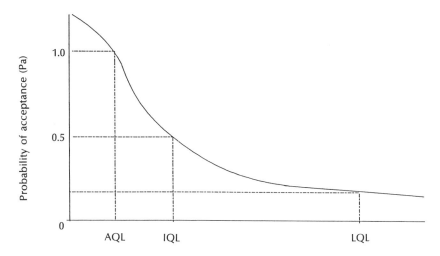

Figure 13.8 Indices used in developing quality sampling plans.

be low; and the indifference quality level (IQL) related to the position somewhere between the AQL and LQL. These relationships are illustrated in Figure 13.8.

13.12 Sampling plans

Sampling plans fall into two categories — variable and attribute. Variable plans have an advantage that, compared to attribute plans having the same designed risk, smaller sample sizes are used. However, each quality characteristic must be measured separately and therefore plotted separately. Attribute plans do not need to do this, as the need for a larger sample results in the ability of treating several measured characteristics as one single operating cluster, and this related to one group of acceptance criteria.

13.12.1 *The single sampling plan*

Here, a random sample (n) is taken from the product lot. If the defect rate is seen to be less than or equal to the acceptance number (c) for the sampling plan, then the lot is accepted. However, if the defect rate indicates a value higher than the acceptance number, then the lot is rejected. Thus, the sampling has an effect on the actual decision to accept or reject the lot.

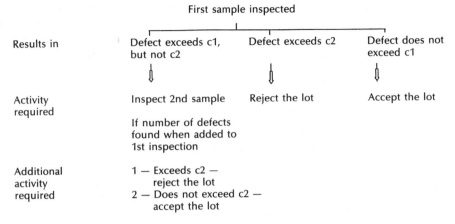

Figure 13.9 Double sampling plans.

13.12.2 *The double sampling plan*

Again, a random sample is drawn (n), which is generally smaller than in the single sampling plan. If the defect rate is seen to be very small compared to the acceptance number (c) for the sampling plan, then the lot is accepted. However, if the defect rate indicates a value much higher than the acceptance number, then another sample is drawn. If the result is conclusive in the second sample (a high or a low result), then either the lot is accepted or rejected. If it is not conclusive, the lot is rejected.

The acceptance numbers in double sampling plans require the development of two levels for c, one that relates to a lower value, and another that relates to a higher, but still acceptable, value. An explanation of this can be seen in Figure 13.9.

13.13 The operating characteristic curve

The operating characteristic (OC) curve is a visual representation of the risk attributed to the producer and the consumer. It uncovers the notion that inspection — whether 100 per cent or sampling — will not guarantee that every defect will be accounted for. Risk seen in terms of the producer (α) relates to non-defective product that will be rejected; and in terms of the consumer (β) of defective product that is accepted.

The OC curve is a graph of the proportion defective in a lot, as compared to the probability that the sampling plan will accept the lot. P is unknown and therefore the probabilities are stated for some uniform value between 0 and 1.

STATISTICAL PROCESS CONTROL 269

The basic premise of the OC curve is to maximize the acceptance of good lots and minimize the acceptance of bad ones. There will, however, always be the case of accepting bad product because the OC curve is an approximation, rather than an absolute representation. OC curves should *not* be used to predict quality of lots submitted or predict the end quality level.

13.13.1 *Constructing the operating characteristic curve*

The OC curve is essentially constructed by determining the probability of acceptance (POA) for a given number of values of incoming defect rate. The POA is the probability that the number of defectives found in the sample is equal to or less than the acceptance number for the sampling plan. Three distributions are used to evaluate the appropriate POA of the OC curve; the binomial, hypergeometric and Poisson distributions.

13.14 Process capability

Process capability is essentially about evaluating whether a process consistently produces product within the specifications set down by the product/service process designer. It is essentially about determining a quantifiable prediction of process performance. Evaluating the process is usually for testing whether a process can produce a given product within designed specifications. It is therefore a tool that is becoming more frequently used in quality planning. Juran and Gryna (1993) indicate that process capability is the measured, inherent variation of the product output from any given process. Capability therefore refers to the ability of a process to deliver product consistently to the specifications.

13.14.1 *Process capability uses*

Uses of process capability include:

1 The provision of data about a process to designers in order to understand the limitations they have to work towards.
2 Choosing the most appropriate process to manufacture a product or service.
3 Ensuring that machines are used in the most appropriate process relevant to their capability.
4 Evaluating sequential processes and evaluating their effectiveness in relation to the overall process requirements.

The wide range of application and the proactive approach of the method provide a basis for an effective tool that seeks to ensure that processes are tested and appropriately assigned according to design needs and requirements.

Process capability is assumed to encompass $\pm 3\sigma$. In this respect, tolerances

Table 13.7 Process capability index (C_p) and specification limits

Process CI	Specification limits (two-sided) %
0.50	13.36
0.67	4.55
1.00	0.30
1.33	64 ppm
1.63	1 ppm
2.00	0

and 6σ are inherently similar and their similarity in context provides the basis for an equitable comparison. It is in this respect that designers can, through understanding working processes in real life, determine the limits of their design creations.

$$C_p = \text{capability ratio} = \text{specification range/process capability} = \frac{US_1 - LS_1}{6\sigma}$$

The capability ratio can be used to provide a measure of the capability of the process. Table 13.7 illustrates the ratios that are generally used. The impact of the ratio means that real processes can be compared with each other and against the standard contained in this table.

General assumptions about the interpretation of the results of a process capability study include:

1   Sufficient data have been collected to provide the basis for a full examination of the process under consideration.
2   That the process is determined as in a state of statistical control.

These assumptions should be tested to ensure their validity in real life. What constitutes sufficient data for one process may not be sufficient for another.

13.15 SPC and quality improvement — what it means

Process control means evaluating a given process, determining the causes of quality-related problems that affect its control and then eradicating them. A distinction must be made between primary and secondary control requirements. Primary control is about bringing a given process to a stable state, in effect in statistical control, where common causes of variation are reduced to a minimum. Secondary control is about improving the process in order to give more enhanced quality control performance. This suggests that processes in statistical control may not be capable of assuring the quality of product and that any quality problems may persist. As Deming (1986) suggests 'removal of a special cause of variation, to move toward statistical control, important though it may be, is not improvement of the process', at least in the short term.

Secondary control means targeting:

1. *Process average* — The process average although correctly calculated may be a result of improper equipment settings, caused either by design or human error. Other factors that can affect the process average include supplier input-related anomalies which were detected, the actual product design and possibly environmental parameters that affect both equipment and human performance characteristics.
2. *Process drift* — This refers to managing the process drift in such a way that the process can be reset in order to provide improvement. However, analysis of process drift cannot be done in isolation; it therefore must be done as an ancillary or addendum to others factors and in this respect is an after-the-fact evaluation.
3. *Cyclical changes in the process* — This is related to process drift, but is a result of a consistent trend of change. At any stage in the cycle, product quality may not result in customer satisfaction. Breaking the cycle means finding the underlying causes in order to reduce their effect.
4. *Measurement techniques* — Some techniques may not provide the consistency required, i.e. techniques may be adequate but poorly applied, or that inappropriate variables are measured.

Process improvement essentially means not only measuring and analyzing the process under consideration, but also the underlying factors that contribute to the consistent delivery of product or services.

This includes both input and in-process variables and may result in changes to the average value, reducing variation and providing a more tightly controlled process, or changing the product/service design specifications. Process improvement also means not being satisfied with the status quo, and searching for ways and means to provide a tighter process that can be relied upon to produce product or services consistently, with minimum cost.

13.16 Process variability reduction

The best result that can be expected from a process is when the centreline equals the designed target specification and the upper and lower limits are also contained on this line. The process is then in perfect control. This ideal state rarely if ever occurs. If the process is seen to be within the tolerances set by the designer, then improvements can still be made to ensure tighter control. For example, if the process variation is 1α or less, and the value in the marketplace is considered relatively low, i.e. low risks in relation to safety or reputation, then a reduced inspection programme could be developed and used. Conversely, if the process variation is greater than 1α, then the programme of inspection should continue and greater effort is needed to reduce the variation to within 1α.

This may seem unnecessary when the process is considered in control. What needs to be understood, however, is that one such process forms part of a chain of processes in the production of a product or service. Consequently, there is a need to ensure that *all* processes move towards the ideal state. It is accepted that some processes in the chain will not be as effective in relation to reduced variation as others, and it is just this that provides the motivation to improve all processes, helping to assure the delivery of designed product or services. This requirement takes on greater relevance when the product or service affects safety, or risks of failure are considerable in terms of cost or time. It could also mean improved product performance and capability and more effective and less costly after-service conditions.

Variability affects the cost of a product and process operation, its capability to deliver a product within specifications and the output operating conditions of the product or service itself. Managing variability is therefore a key feature in managing the quality-related characteristics of products and services to satisfy customers requirements consistently.

13.17 Taguchi — the quality loss function

The quality loss function directly relates economics to variability. Any deviation from the ideal means that there is wastage, so variation costs. What is seen as inappropriate today is the concept that economic loss occurs *only* when product or services do not meet specifications. The quality loss function can be quantified as the loss due to variation by using the quadratic equation:

$$L = k(X - T)^2$$

Where L = Loss in terms of money.
k = Cost coefficient.
X = Value of quality characteristic.
T = Target value.

The quality loss function may actually provide more administrative costs than customer-perceived benefits.

Thus, there is a limit to the application of this mathematical technique to managing variation in processes, especially in relation to reducing the costs of wastage. A convincing methodology in its own right, it has the ability to reduce costs, but to a point, and it is just this point that is difficult to determine. The costs of manipulating and managing the production process may be too great the nearer the variation gets to the ideal state. Consequently, quality managers choose to minimize overall or total cost of the process, rather than just to concentrate on the costs associated with reducing variation. This means that resources can be given to other processes that have greater concerns in relation to quality problems.

13.18 Chapter review

Statistical process control (SPC) is the application of statistical techniques to a process in order to develop statistics about that process and apply control techniques to manage that process. It involves the measurement of data surrounding variation in a process. Statistical techniques in themselves do not provide the quality of product demanded by customers, but do provide a methodical means of assuring that designed specifications are consistently met.

Statistical control charts are a graphical means of displaying the variation over time of an examined process. They provide a means by which past data can be used to provide limits — upper, centre and lower — of the examined process, in terms of variation. The limits could be designed (tolerances) or they could be calculated from the data generated out of the process. An evaluation of the graph can indicate whether the sample is common (random non-assignable), or has a special cause (assignable) which can be managed and minimized.

Control charts can be divided into two main types — variable (objective) or attribute (more subjective). Each can be applied in differing circumstances depending upon the type of data characteristic to be generated.

CuSum charts are used to detect changes or shifts in the sample mean of a process. They are interpreted by comparing plotted points with defined limits. The main advantage over Shewhart control charts is that they consider all information from the observations up to and including the point of evaluation.

Acceptance sampling involves evaluating a portion of a product grouping in order to make a decision about whether to accept or reject an assignment of product. The major reason for developing this technique is to reduce cost, time and reduce product destruction. Sampling alternatives range from no sampling to 100 per cent examination. Acceptance plans are designed to reduce the risk to the manufacturer of passing a defective product and reduce the risk of a product being accepted by a customer. Appropriate management of the sampling plan will ensure that an effective balance is achieved between these two competing criteria.

Process capability is essentially about evaluating whether a process consistently produces product within designed specifications. It can be used to provide reinforcement of the present state of performance of a process, or it can be used to provide a planning tool in order to test known tolerances against the process under examination.

Process control means evaluating a given process, determining the causes of quality-related problems that affect its control, and eradicating them. Primary control (bringing a process into a stable state) and secondary control (improving that process) are techniques that are consistent with the quality improvement philosophy.

The quality loss function directly relates economics to process variability, where any deviation from the designed target means wastage. Its main disadvantage seems to be that it can develop greater administrative costs than the perceived customer benefits of tighter targeting and thus distracts from using resources to secure greater quality-related problems elsewhere in the organization.

13.19 Chapter questions

1. Define the term SPC.
2. Evaluate the process of inspection.
3. Discuss statistical process charts.
4. Discuss quality (CuSum) control charts.
5. Evaluate acceptance sampling techniques.
6. Evaluate the impact of acceptance plans.
7. Describe sampling plans.
8. Evaluate the operating characteristic curve.
9. Describe process capability.
10. Evaluate the cumulative summation (CuSum) control chart.
11. Compare the Shewhart and CuSum charts.
12. Evaluate the relationship between SPC and quality improvement.
13. Discuss the implications of Taguchi's quality loss function.

CHAPTER 14

Quality economics

CHAPTER OBJECTIVES

Explain the basis and importance of quality-related costs
Explain why it is necessary to measure costs
Compare and contrast the cost of quality versus the cost of non-quality
Explain the hidden costs of quality
Describe the impacts of lifecycle costs
Discuss how to manage quality costs effectively

CHAPTER OUTLINE

14.1   Introduction
14.2   What are quality-related costs?
14.3   Classification of quality costs
14.4   The importance of quality costs to the quality-oriented organization
14.5   Quality costs — why measure them?
14.6   Cost of quality versus cost of non-quality
14.7   Hidden costs of quality
14.8   Lifecycle costs
14.9   The management of quality costs
14.10 Chapter review
14.11 Chapter questions

14.1 Introduction

All organizations to some extent use financial controls to manage themselves. The degree generally differs in relation to the size of the organization and whether it is a private or a public company. Nonetheless, organizations use cost and financial data in order to operate their concern. Montgomery (1985) indicates that there are several reasons why 'the cost of quality should be explicitly considered in an organisation':

1. An increase in the cost of quality because of increases in technology use.
2. Increasing sophistication of end-users in their consideration of lifecycle costs.
3. Internally, an increase of the use of quality cost data to make quality management-related decisions.

14.2 What are quality-related costs?

The definitions provided in British Standards Institution standards BS 4778, BS 6143 and BS 5750 or BS EN ISO 9000 (see Chapter 15), illustrate that there is no universal consensus upon just what constitutes quality costs. Traditionally, costs of quality (COQ) have been considered as the cost of running a quality assurance system (complete or in development), with perhaps the inclusion of other costs, such as scrap and warranty costs.

Present thinking would now tend towards accepting that quality costs are incurred in the design, implementation, maintenance and improvement of a quality system. Quality costs can be attributed to many work activities in an organization. In this regard, COQ crosses inter- and intradepartmental boundaries, much as the process for developing and producing the product or service does in that organization. No department or group is isolated and therefore is not immune to the constraints and opportunities of managing quality costs. Managing costs means managing the COQ, as quality is a company-wide activity. It is also conceded that COQ is not confined to the internal environment of the organization, as the activities of, say, suppliers, etc., can affect the outcomes of costs related to quality.

14.3 Classification of quality costs

This discussion will focus primarily on BS 4778, as this standard appears to be one of the most widely used in this context. BS 4778 — Part 2 classifies quality costs into the following:

12.3 *Quality related costs*. The expenditure incurred in defect prevention and appraisal activities plus the losses due to internal and external failure.
12.4 *Prevention costs*. The cost of any action taken to investigate, prevent or reduce defects and failures. Prevention costs can include the cost of planning, setting up and maintaining the quality control system. They also include process design, product and service design and employee training schemes.

12.5 *Appraisal costs*. The cost of assessing the quality achieved. Appraisal costs can include the cost of inspecting, testing, etc. carried out during and on completion of manufacture of product or service.

12.6 *Failure costs* — internal. The costs arising within the manufacturing processes of the organization of the failure to achieve specified quality. This can include the cost of scrap, rework and re-inspection.

12.7 *Failure costs* — external. The costs arising from outside the manufacturing organization of the failure to achieve quality specified. The term can include the costs of claims against warranty, replacement and consequential losses of custom and goodwill.

Quality costing is one of the most important of the many tools and techniques available to management in their search for managing in a quality way. Indeed quality costing is seen as a primary measure when applying TQM. Quality costing should not be considered an end in itself, but should be seen as a means to quantify quality activities in a language that top management and possibly shareholders can more readily understand. The success of the quality system depends upon many other inputs and activities. Dale and Plunkett (1991) indicate that 'a knowledge of quality costs helps managers to justify the investment in quality improvement and assists them in monitoring the effectiveness of the efforts made'.

Quality costing appeals to top management, as its language is in tangible terms about financial entities. Using quality costs in this way for developing and encouraging quality improvement will substantiate the application of TQM philosophies and practices.

14.4 The importance of quality costs to the quality-oriented organization

The National Economic Development Office claimed that 10–20 per cent of an organization's total sales value could be accounted in quality-related costs. This is further supported by Dale and Plunkett (1991) in that 'quality-related costs commonly range from 5 to 20% of company annual sales turnover'. Generally, 95 per cent of the total quality-related costs are expended on appraisal and failure elements. Failure costs must be regarded as avoidable and a reduction in such costs is usually attributable to such activities as eliminating causes of non-conformance, which may lead to a reduction in appraisal costs. There is therefore a demonstrated need to balance future failure costs with appraisal costs. However, costs of failure cannot always be accounted for, i.e. how do you measure or effectively appraise customer satisfaction? Managerial requirements suggest that what can be measured, can be managed, a Deming byline. This means that appraisal costs will increase and companies will therefore spend financial resources in order to gain by reducing failure costs.

Robertson (1971) states that for the 'average UK organisation' (whatever this means) the analysis of quality-related costs are 65 per cent failure costs, 30 per cent appraisal costs and 5 per cent prevention costs, and further indicates that quality-related costs may be in the range of 4−20 per cent of sales turnover. These figures are corroborated by Abed and Dale (1987), who suggest that as a percentage of total quality costs 67 per cent are failure costs, 28 per cent are appraisal costs and 5 per cent are prevention costs, with total quality costs as a percentage of sales turnover at an average of 9.2 per cent, the range being from 2 to 25 per cent. A further example comes from Garvin (1983), who compares Japanese air-conditioning manufacturers with their American counterparts, focusing upon warranty claims. Garvin indicates that Japanese company warranty claims are about 0.6 per cent of sales turnover, the best an American company could report was 1.8 per cent, with the worst being 5.2 per cent. In comparison, Dale and Plunkett (1991) illustrated the results of examples of quality costs experienced by several British companies:

> British Aerospace Dynamics — quality costs are 11% of the total cost of production.
> British Airways Technical Workshops — staff time on quality-related matters are — Failure (22.9%), Prevention (19.4%) and Appraisal (6.8%).

14.5 Quality costs — why measure them?

Measuring quality costs will provide a means to quantify in management terms the effect that quality-related activities have on organizational performance. This is especially so in the areas of production, marketing, supplies procurement and customer satisfaction. It should positively influence employees and their attitudes towards the quality system, TQM and related continuous quality improvement schemes and practices.

Measurement of quality costs will focus attention upon such areas as appraisal, prevention and failure and therefore provide opportunities for cost reductions. There is a danger that quality costs may, due to the powerful language, become the focus, rather than the quality activities they should be measuring. Management would be failing their own customers, employees and their organization as a whole if this were to become a reality. Sadly, for many organizations, this *has* become the reality.

Performances across a wide range of quality-related activities may need to be measured and this will provide a basis for internal quality cost comparisons between departments, processes, services and products. The measurement of quality costs can be clearly seen as a major step towards quality control, quality improvement and TQM.

14.6 Cost of quality versus cost of non-quality

Cost of quality is the understanding by management of the savings that can be made by introducing TQM and in particular the relative small cost of quality versus the ongoing cost of non-quality. This will include the cost of implementation of the system and the ongoing administration.

The cost of quality can be divided into three main aspects:

1. Failure costs.
2. Appraisal costs.
3. Prevention costs.

From these, only prevention can be regarded as a cost of quality, whereas the other two are in essence the cost of non-quality — inspection and rework of errors; rather than the principle of working to attain zero defects. The terms used here may sound more like those used for the manufacturing industry. However, if service is regarded as the end-product of a process, this statement is equally valid for any customer service encounter.

Albrecht and Zemke (1985), in reference to the Technical Assistance Research Programs Inc. findings, show that the cost of non-quality in the service industry may be very high.

The intangible nature of the service product means that the mistakes, which can never be undone, will have to be regarded as a cost factor. Clients, who have been dissatisfied may not only never return, but may even influence other people into not using the defective services that the organization provides. What losses are incurred because of this process are very difficult if not impossible to measure. It seems that because they are difficult to substantiate, many organizations have turned a blind eye to these losses.

However, if they are truly committed to TQM, it will be recognized that one aspect able to reduce this cost of non-quality is to ensure that the incidences are reduced to an absolute minimum. This can be accomplished through proper training of staff in the communications skills required to be able to assess clients, needs better and to bring customer satisfaction to the optimum — whether internal or external customers.

14.7 Hidden costs of quality

Where errors in manufacturing produce waste, scrap or rework, then the hidden cost of quality or rather non-quality can be seen as:

1. The extra material needed to be supplied to accommodate this extra wastage.
2. The extra manpower costs of labour and perhaps overtime.

3   The opportunity cost of working on a part the second time round or, in the case of a scrapped item, on a completely new part.
4   Possible delays in the ultimate shipment of the order.
5   Increased machine maintenance and repair costs.
6   Increased risk of machine breakdown.
7   Reduced production capacity resulting from the need to overproduce in order to manufacture a given quantity of production items.

The hidden costs of quality are a financial, human and physical erroneous liability that can be addressed effectively through the application of TQM practices.

14.8 Lifecycle costs

Juran and Gryna (1993) discuss the impact of the lifecycle cost theory. All products/markets/services have lifecycles. Fashion, for example, is a cycle. What is interesting about Juran and Gryna's application is that the cost of the product/service should not just be limited to the cost at purchase. It should also include the cost of maintenance and the running cost of the product.

Designing a product that lowers the overall lifecycle cost may mean that the initial cost may be higher than originally anticipated, but the consumer would benefit in the long run. An example is that of laser printers. For some printers, the purchase cost is far exceeded by the actual running costs. In others, the cost of after-sales — not including running costs — may again exceed the original purchase price. Gryna (1977) discussed the concept of user failure cost — that cost to the consumer for failures of the product over the life duration of that product. This is difficult to apply generally, as the forecasting of specific failure costs to the consumer is impossible. However, anticipation, based on previous experience — as in the case of mass producers — would certainly provide a basis for the analysis. Marketing opportunities such as this have been used by world-wide companies. An example is that of Kyocera Electronics, who advertised in the November 1994 edition of *Personal Computer World* that the cost of their laser printer — over a three year period — was not the cheapest to purchase, but the cheapest to run over that time in costs/page printed. This is a part of quality costs that needs to be brought out into the open.

14.9 The management of quality costs

Surveys conducted by Roche (1981) and Duncalfe and Dale (1985) suggested that only about one-third of the companies studied actually collected quality cost data and that these findings indicated that less than 40 per cent of companies collect and analyze quality cost data in a *systematic manner*. The costs most measured were suggested to be those for cost of scrap, rework and warranty claims.

Although Tsiakals (1983) highlighted the importance of not placing complete reliance upon the data from quality costing as a means of improving quality and reducing costs, quality cost data should be used as some basis for the quantification of quality-related activities, but not solely to be used as a weapon by top management to cut costs, as mentioned previously.

Quality costs analysis therefore should be the backbone of a quality costing system. It is here that systematic collection, analysis and reporting will provide the necessary information for effective quality-related decisions to be made. But where should you start? Using the seven old tools of quality in this respect can provide a basis for complete and effective analysis of quality costings.

For example, applying Pareto analysis to a production line will result in the discovery not only of the actual problem, but of the cost of that problem to the organization. It will also serve to provide the basis for savings that could be achieved if the highlighted problems were eradicated. In this sense, quality cost analysis provides for the integrated use of a variety of quality problem-solving tools.

Whatever the outcome of the quality cost analysis, particular attention must be paid to solving the highlighted problems. Throwing vital resources into appraisal rather than prevention is not good practice, as many organizations have found to their folly. It is just this point that has made Japanese manufacturers more effective than their American and European counterparts during the 1960s and 1970s. Americans and Europeans funded appraisal rather than prevention; they were essentially targeting the *symptom* rather than the Japanese approach of targeting the *core* problem and developing an effective solution to it.

Quality cost data must be developed from the real outcomes of quality activities, but rarely do you find the data available to you in a form that is usable. More often, quality cost data have to be developed after a specific quality cost system has been developed. The message here is that serious attention must be given to planning the needs and outcomes of the quality cost data — not only in its analysis, but also in the use the data will be put to. Where an organization has a quality assurance system or better a quality management system (BS EN ISO 9000), then related quality costs are available according to the system requirements and level of development.

Quality cost data should not be used as a means to control costs, but as a means to improve the quality of products produced. This has long been the unstated rule and one that typifies why some quality cost programmes fail. They fail because management objectives are not in line with improvement requirements and they get disillusioned when the system does not perform as expected. Reduced costs do not specifically come from increased quality, e.g. as the appraisal costs increase, costs of failure reduce, but there is a time when failure occurs no matter what appraisal costs are incurred. Overemphasis on quality cost reduction, geared to short termism through the accounting mechanism should not be used.

Management must manage quality — not administer — as far as quality costs are concerned.

14.10 Chapter review

The traditional approach to quality costs can be seen as the costs attributed to running a quality assurance system, with perhaps the costs associated with scrap and warranties. This has been superseded today, although a consensus of what quality costs is demonstrably weak. Various definitions contained in BS 4778, BS EN ISO 9000 and BS 6143 indicate the problem.

The importance of managing quality costs cannot be overemphasized, especially when quality-related costs can account for as much as 5–20 per cent of annual turnover. Managing these quality costs down to a minimum could in these instances mean a reduction in price, the gain of the whole market share and an increase in overall profits. Measuring quality costs is important because it quantifies in managerial terms the effect and impact of quality-related activities. Measuring quality costs will focus attention upon such areas as appraisal, prevention and failure. It will also provide a means for tracking quality cost performance over time.

Quality costs can therefore be classified (BS 4778) into prevention, appraisal and internal and external failure costs.

A number of studies have indicated that fewer than 40 per cent of organizations deliberately and systematically collect quality-related cost data. However, quality costs data should be used to provide a basis for continual improvement.

14.11 Chapter questions

1 Explain the basis and importance of quality-related costs.
2 Explain why it is necessary to measure costs.
3 Compare and contrast the cost of quality versus the cost of non-quality.
4 Explain the hidden costs of quality.
5 Describe the impacts of lifecycle costs.
6 Discuss how to manage quality costs effectively.

CHAPTER 15

Quality standards

CHAPTER OBJECTIVES

Explain the meaning of a quality system
Evaluate the basis, content and benefits of using BS EN ISO 9000 in organizations
Explain the processes of certification and accreditation
Explain the process and necessity of auditing
Evaluate the impacts of other standards that may be consulted

CHAPTER OUTLINE

- **15.1** Introduction
- **15.2** What is a quality system?
- **15.3** BS EN ISO 9000
  - 15.3.1 *Origin and basis*
  - 15.3.2 *The parts of BS EN ISO 9000*
  - 15.3.3 *BS EN ISO 9000 — update 1994*
    - General
    - Specific changes and implications
- **15.4** Certification and accreditation to BS EN ISO 9000
  - 15.4.1 *Definitions*
  - 15.4.2 *Certification*
    - First party assessment
    - Second party assessment
    - Third party assessment
  - 15.4.3 *Benefits of certification to BS EN ISO 9000*
    - General audit process for certification
  - 15.4.4 *Accreditation*
    - NACCB — National Accreditation Council for Certification Bodies
    - Certification — scope
    - Accreditation — scope of a certificating body
    - The process of accreditation

- 15.5 Auditing to BS EN ISO 9000
- 15.6 BS 7850 — total quality management
- 15.7 Other standards that may be useful to consult
  - 15.7.1 *BS 7750 — environmental management standard*
  - 15.7.2 *Investors in People (IIP)*
  - 15.7.3 *European Quality Award (EQA)*
- 15.8 Implications of the application of BS EN ISO 9000 — a service approach — higher education
- 15.9 Chapter review
- 15.10 Chapter questions
  - Appendix 1 — BS EN ISO 9000 Quality System Elements
  - Appendix 2 — BS EN ISO 9000 Survey Instrument
  - Appendix 3 — EN 45012 Clauses

15.1 Introduction

Quality management standards are seen as a major pillar supporting the drive for continuous quality improvement through TQM. In this chapter we explore two different but related quality standards. These two standards are:

BS EN ISO 9000 — Quality Management System
BS 7850   — Total Quality Management Standard

Another standard that may be useful is BS 7750. Investors in People and the European Quality Award (although these are not technically quality standards in the same vein as BS EN ISO 9000) may also provide some basis for a structured approach to quality improvement practices, as well as gaining external recognition. These are discussed at the end of the chapter.

15.2 What is a quality system?

The quality system is designed to provide both the support and mechanism for the effective conduct of quality-related activities in an organization. It is a systematic means to manage quality in an organization. The quality-oriented organization ensures that a quality management system is in place and working effectively.

15.3 BS EN ISO 9000

Here, BS EN ISO 9000 is seen as a voluntary management standard and it is accepted that this standard forms a major pillar supporting the development and operation of TQM in an organization. BS EN ISO 9000 is not a standard to manage products as such, as it is management-process-oriented. This means that you cannot use products in themselves as a basis for the demonstration of adherence to the BS EN ISO 9000 standard. This can only be done, on-site, and through the accepted process of quality auditing. Defective products in themselves do not provide *prima facie* evidence of non-conformities in the quality system. In fact, 100 per cent product defects may not of themselves indicate that the quality system itself is not operating effectively. BS EN ISO 9000 therefore bears no relationship to the product process output, but rather to the management process outputs. It is conceivable that if an organization wants to produce supposedly defective product — and does it consistently, that the use of BS EN ISO 9000 in these circumstances would be acceptable. It depends on what management want the quality management system to do.

Increasingly organizations certificated to the standard ensure that they are supplied by an organization which is also certificated to the standard. This creates an overwhelming pressure for suppliers — especially to government departments — to become certificated, whether they want to or not. This aspect provides some basis for the notion that certification to BS EN ISO 9000 does not mean that the organization is quality-oriented. It could mean that in order to survive, the quality management standard is used.

BS EN ISO 9000 is used by some organizations purely as a marketing exercise. Quality is given a cursory evaluation. Its main use to them is to indicate to present and prospective customers their supposed commitment to quality. It is a PR exercise and on many occasions it fails. The problem with this strategy is that customers do not then believe that the standard is worth much, because of the demonstrated failures, and this creates problems for those organizations committed to quality.

15.3.1 *Origin and basis*

Commercially oriented quality system standards have evolved over the past thirty years or so. Much of the initial development was for major military projects in the United States. For this, MIL-Q-9858 was developed. Table 15.1 indicates some of the development to date.

The basis for the development of BS EN ISO 9000 can be squarely placed on its military background. The change came about in 1984, when the British Standards Institution managed to persuade the International Standards Organisation (in Switzerland) to develop a generic quality management standard for world-wide use, based on its revised standard of 1979.

Up to twenty-six countries were initially involved and all produced clones in 1987, when the new BS EN ISO 9000 standard was finally published. The result depicted a minimum standard that was acceptable to the countries involved. They essentially differed in language, forewording, numbering and title. Nevertheless,

Table 15.1 Quality standards development

Quality system standard	Year	Reason	Country
MIL-Q-9858	1963	Military	USA
AQAP	1969	Military	NATO
10 CFR 50	1970	Nuclear federal regulations	USA
ANSI-N45-2	1971	Generic quality enhancement	USA
DEFSTAN 05/**	1973	Military	UK
CSA Z299	1975	Generic quality enhancement	Canada
AS 1821/2/3	1975	Generic quality enhancement	Australia
BS 5750	1979	Generic quality enhancement	UK
ISO 9000	1987	Generic quality enhancement	International
National and European clones	1987		
BS EN ISO 9000 (Review)	1994	Generic quality enhancement	International

the published standard represented a great achievement in international co-operation, at least for quality.

In 1987, the world's quality management standards were amalgamated into the BS EN ISO 9000 series of standards — and as such provide 'the refinement of all the most practical and generally acceptable principles of quality systems'. In this respect, BS EN ISO 9000 was designed to replace national standards and help international trade by providing a quality management system that was equitably based on international requirements. The BS EN ISO 9000 is a series of standards designed to provide a guide to the systematic operation of quality-related activities in an organization.

Since 1987, many military standards themselves have given way to BS EN ISO 9000. In the United Kingdom at least, the Ministry of Defence (MOD) has almost dispensed with the requirements of AQAP and now insists on suppliers using BS EN ISO 9000. The move comes from the third party requirements of BS EN ISO 9000 as compared with AQAP, which necessarily involved second party assessments.

In this text we will confine ourselves to the specific application of the BS EN ISO 9000 quality management standard. However, Figure 15.1 indicates the

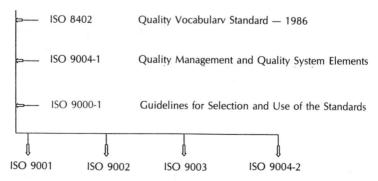

Figure 15.1 BS EN ISO 9000 series of quality-related standards.

relationship of BS EN ISO 9000 as a standard to the content of the BS EN ISO 9000 series of standards. BS EN ISO 9000-3 is considered here, but refers to the application of ISO 9001 to the development, supply and maintenance of software.

15.3.2 *The parts of BS EN ISO 9000*

BS EN ISO 9000 is broken into four parts. This provides a means for targeting the actual standard application after consulting the guidelines in ISO 9004. These parts are:

1. *ISO 9001* — Model for quality assurance in design/development, production, installation and servicing.

    This standard is for demonstrating to customers the derived quality of the quality management system for all quality-related processes from design through to after-sales service. It is the *highest* level in terms of the depth of procedural coverage.

2. *ISO 9002* — Model for quality assurance in production and installation.

    This standard provides a demonstration of quality-related processes that affect activities in production and installation. It does not provide a demonstration of design elements. It is assumed that the design quality is already demonstrated and proved elsewhere.

3. *ISO 9003* — Model for quality assurance in final inspection and testing.

    This standard applies to the contractual relationship between supplier and customer. Product quality in relation to the manufacturing process is assumed to be the responsibility of the organization's supplier. Consequently, there is a much reduced depth of standard content.

4. *ISO 9004* — Quality management and quality system elements — guidelines

    A primary purpose of the guidelines is the provision of an 'adequate description of the quality management system while serving as a permanent reference in the implementation and maintenance of that system'. This is a guideline to help current or prospective users of BS EN ISO 9000 to choose the correct part of the standard for their specific purposes. Nevertheless, an organization should try to develop a quality management system that will suit their requirements as well as their customers'. For example, customers may actually require a *higher* level than the organization would like to implement. The commitment to quality therefore must be demonstrated, which creates a greater impetus for developing a quality management system to satisfy that customer.

The sections, title and their corresponding application can be seen in Table 15.2.

BS EN ISO 9000 quality system elements are contained in Appendix 1, at the end of this chapter. What is interesting here is that the clauses are numbered

Table 15.2 Sections of BS EN ISO 9000

Title	Corresponding section — BS EN ISO 9000			
	9001	9002	9003	9004
Management responsibility	4.1+	4.1*	4.1#	4
Quality system	4.2+	4.2+	4.2*	5
Contract review	4.3+	4.3+	—	7
Design control	4.4+	—	—	8
Document control	4.5+	4.4+	4.3*	17
Purchasing	4.6+	4.5+	—	9
Purchaser-supplied product	4.7+	4.6+	—	—
Product identification and traceability	4.8+	4.7+	4.4*	11.2
Process control	4.9+	4.8+	—	10
Inspection and testing	4.10+	4.9+	4.5*	12
Inspection, measuring and test equipment	4.11+	4.10+	4.6*	13
Inspection and test status	4.12+	4.11+	4.7*	11.7
Control of non-conforming product	4.13+	4.12+	4.8*	14
Corrective action	4.14+	4.13+	—	15
Handling, storage, packaging and delivery	4.15+	4.14+	4.9*	16
Quality records	4.16+	4.15+	4.10*	17.3
Internal auditing (QS)	4.17+	4.16*	—	5.4
Training	4.18+	4.17*	4.11#	18
After-sales servicing	4.19+	—	—	16.2
Statistical techniques	4.20+	4.18+	4.12*	20
Economics of quality	—	—	—	6

+ Full requirement.
* Less stringent application than BS EN ISO 9001.
# Less stringent application than BS EN ISO 9002.

differently between the various levels ISO 9001–9003. This can be confusing. Also, the numbered clauses in ISO 9004 do not directly correspond with the numbered clauses in ISO 9001–9003. The review in 1994 does not appear to address these issues. However, Table 15.2 may help.

Other differences include:

1. ISO 9002 in comparison to ISO 9001:
   (a) No design reviews.
   (b) No management responsibility for test and monitoring of design elements.
   (c) No design control.
   (d) Limited training requirements focused on production and installation.
   (e) No servicing requirements.

2. ISO 9003 in comparison to ISO 9002:
   (a) Ensuring that the quality policy is understood by all is not required.
   (b) The QMS is limited to inspection and testing.
   (c) Traceability requirements are excluded.

(d) Internal audits are not required.
(e) Non-conformances are not required to be recorded at all.
(f) Document control mechanisms are not required.
(g) Quality records are limited to product records.
(h) Training requirements do not include training analysis evaluations.

ISO 9003 is a very limited standard, and its usefulness to a quality-oriented organization is therefore severely limited. In New Zealand, for example, ISO 9003 is seen as a small business application. However, more than 95 per cent of the certificated organizations are certificated against the ISO 9001/2 standards.

A BS EN ISO 9000 survey highlighting the perceived importance given to each section is included in Appendix 2 at the end of this chapter. This may be of use in the determination of the state of the quality culture of a department or an organization prior to and during the implementation of BS EN ISO 9000.

15.3.3 *BS EN ISO 9000 — update 1994*

The revision of BS EN ISO 9000 has come about after much debate and deliberation over the derived inadequacies of the original 1987 standard. The argument for service industry adoptions of the use of BS EN ISO 9000 was that it laid emphasis on the development of bureaucratic and paper-driven systems, rather than creating a culture that provided for the application of flexibility and innovation.

Some changes are cosmetic (e.g. changes to text), but others have much more of an impact on the effective development and use of a quality system. The update is considered to be a significant improvement on the 1987 version, as many implied requirements of that version are now more clearly defined and subsequently now need to be addressed. Also, previous facets that were advisory are now required to be assessed. However, the major criticism of the 1987 version of bureaucratization has been addressed by the updated standard. In this respect, the standard states that 'it is not the purpose to enforce uniformity of quality systems'. However, a standard, by its very nature and application, is about this. What the updated standard attempts is the rationalization of the applied process, rather than the content of the requirements of the standard. The new standard tries to nullify this by providing an organization with the capability of choosing its own quality management system and its subsequent requirements.

General

- *Quality policy* — A major change in the new version is that external parties are considered multi-customers rather than just one customer–supplier relationship. The quality policy needs to be 'relevant to the organisational goals of the supplier and the expectations of customer(s)'. Relevant means developing quality practices and procedures that are useful and necessary to the organization in order to help fulfil those quality goals.

It does also mean that small organizations can target their quality system requirements more appropriately. The emphasis given to 'management representative shall have executive responsibility for quality' suggests greater consideration is given to demonstrate the quality-related commitment of top management. It also includes the responsibilities for internal (quality system or elements) and external bodies (certification body).

- *Quality planning* — Clause 4.2 now has three subclauses. Clause 1 requires a quality manual rather than just a documented quality system. Clause 2 indicates that the 'degree of documentation required . . . shall be dependent upon the methods used, skills needed and training required'. This means that the documented system needs to be written for those essential procedures. Clause 3 supports the previous version's guidance notes. However, they are now requirements under the 1994 version (rather than just being aware) and therefore it is now necessary to provide proof of compliance.

- *Preventive Action* — Clause 4.14 provides the requirement to take action before a problem occurs. In this regard, the clause forces suppliers to evaluate products and processes, develop effective data about them and take action against any potential non-conformities. The inspection requirements have thus been replaced with a proactive requirement. This is seen as forcing suppliers continuously to reassure their products/services through this evaluation. However, 'actions taken shall be to a degree appropriate to the magnitude of problems and commensurate to the risks encountered'. Risk assessment may become a means for the supplier to defend against the development of continuous data about a given process.

Specific changes and implications

4.1 Management responsibility, management review and time interval of review must be stated. The review meeting agenda must contain a tabled account of the developed quality mission and policy, and the set quality objectives.

4.3 Contract review — an additional *tender* and *order* have been added to the *contract* terms. Implicit here is the need to demonstrate (not assume) that an order can be fulfilled at the time it is ordered. Another important change is that communication channels need to be defined, i.e. contacts between supplier and customer need to be defined explicitly.

4.4 Design and development plans will be defined in terms of the individual responsible and of design and development activities resulting therefrom. In essence, this means that a plan must contain *who* is responsible for a given design and *what* activities are being performed, using *what* resources and *when*. Also new to this version is the requirement to include 'applicable statutory and regulatory requirements'. Furthermore, the contract review needs to account for modification to customer design changes and demonstrate these. Verification also needs to be carried out at 'appropriate stages in the design', suggesting that more than one verification needs to be demonstrated.

4.5 There is now a requirement to control data, and obsolete documents should be identified as such.

4.6 There is the requirement to evaluate contractors, rather than the previous version of assessment of contractors. This has implications for the control of quality audits and records and the maintenance of quality records of subcontractors.

4.9 Maintenance of equipment to ensure process capability is now a requirement. This is seen as the minimum. However, maintenance in relation to other requirements, such as health and safety, may be excluded. This is a matter for consideration.

4.10 In-process inspection and testing paragraph (4.10.3) has been deleted from the new version, as the content is covered in other parts of the updated standard. Also, a clear definition of non-conformance must be developed and implemented — which results in a requirement in the event of a failure to pass or test a product — the procedures for the control of non-conforming product shall apply.

4.14 This is a major revision, with its subsequent increased impact which is discussed above.

4.15 A new requirement is to furnish evidence of consideration of the need for preservation of manufactured product.

4.16 Only addition here is a note that records may be in the form of paper, electronic or other media.

4.17 A requirement to record the implementation and effectiveness of any corrective action that may be verified during follow-up audits.

4.20 The new requirement is to identify all operational areas where statistical techniques may be used. This requires an organization to evaluate the most appropriate statistical methods to use in each business process if any. It also means the review of these operational areas, which includes the process, product and quality management system performances.

15.4 Certification and accreditation to BS EN ISO 9000

Peter Lilley, Secretary of State for Trade and Industry, said in the Department of Trade and Industry (DTI) (UK) Register:

> I believe that increasingly customers are going to look for this assurance of quality before they buy any product or service. BS5750 (BS EN ISO 9000) will become more and more essential for any firm that wants to compete at home or abroad; ... but if the standard is to be of any value,

those who assess quality management systems and award BS5750 certifications have themselves to be the subject of rigorous scrutiny. We have to be sure of their independence, competence and integrity.

This is the subject of this part of this chapter.

15.4.1 Definitions

- *Certification* — This is where 1st, 2nd and 3rd parties evaluate a company's quality system against some specified or stated standard or manual.
- *Accreditation* — This is the mechanism for ensuring the standards of 3rd party certification bodies. The body's certification system is itself evaluated against a certification standard. One such standard is the EN 45011/12 standard.
- *Certification body* — An impartial body, either governmental or non-governmental, possessing the necessary competence and reliability to operate a certification system, and in which the interests of all parties concerned with the functioning of the system are represented.

15.4.2 Certification

A quality system evaluation can take three forms:

1. An organization can assess itself (1st party).
2. Another organization can assess another organization's quality system, and if a customer, it is a 2nd party certification.
3. 3rd party assessments are handled by an independent organization of quality assessors, again by evaluating the organization's quality system against a given quality standard, e.g. BS EN ISO 9000.

A discussion of each of the above follows

First party assessment

Prior to 2nd or 3rd party certification, the organization's quality policy, procedures and detailed work-related instructions are written down and published in the organization's quality manual. A requirement of certification is that the organization must ensure that its quality system complies with these procedures and work-related instructions, i.e. the actual behaviour of individuals within the organization must be reflected in the quality system documentation. This reduces the likelihood of a non-compliance becoming evident during 2nd or 3rd party assessment. This process is known as first (1st) party audit or internal assessment. This occurs in both pre- and post-system certification. The presystem certification or self-audit is a must for any organization seeking to gain 2nd or 3rd party certification. Thus, the self-audit provides a track record of quality system diagnosis, and the basis for quality system improvement and documented achievement.

The problem with self-audit is that it requires an objective measure to translate equitably the quality-related practices of the organization into the prescriptive requirements of the adopted quality standard. It must therefore be managed effectively using control measures to ensure speed of problem solving and learning, and delivery of product and service.

This is particularly true in the case of organizations that are just *testing* the organization's procedures against the adopted quality management standard as a prerequisite to gaining operational experience of managing to those adopted quality standards. Where an organization must develop a track record for the purposes of providing a basis of managing a quality system, then the self-audit will need to be formalized. This formalization will generally correspond with the appointment of an outside lead assessor — assuming the organization does not have such a person in its ranks — to provide legitimacy for the audit and leadership for individuals to experience and learn about the audit process requirements and the developed objective quality outputs.

First party audits should be carried out by staff that have been trained effectively in the skills of quality audits. The results of the audits are fed back into the quality management review process, but this does not mean quality improvement. It just means the development of quality procedures that relate to the translated meanings of the adopted quality management standard.

A quality management review must take place at least once a year and records of the audit and reviews must be retained with other quality records, assuming the adopted quality management standard requires this. The quality reviews should be seen as forming part of the process of continuous improvement. If internal audits locate a non-compliance, then changes in procedures, work instructions and/or re-education may need to be addressed and implemented.

Second party assessment

If an external customer makes an assessment of a supplier against their own or a (inter)national standard, then this is classed as second (2nd) party assessment. The supplier is registered as being in compliance with a quality management system and may be issued with a certificate of registration. Most 2nd party organizations totally re-evaluate supplier's systems every three years. There is a trend towards 2nd party bodies requiring independent registration. Second party assessment provided Britain's first formal quality certification scheme in 1973. Since 1973, it has been UK Ministry of Defence policy that direct contracts for anything other than minor defence equipment are only placed with suppliers assessed against one of the allied quality standards, e.g. AQAP 1, 4, 9 or 13, where AQAP 1 includes AQAP 4, AQAP 4 includes AQAP 9 and where AQAP 13 is always used with AQAP 1. AQAP originated in America, but is not now used.

As from 1 September 1991, MOD ceased assessing companies and required

Table 15.3 Certification in Europe

UK	25 000
Netherlands	1 500
Germany	540
France	1 300
Ireland	860
Italy	800
Belgium	400
Spain	250
Portugal	65
Greece	30

With over 170 000 organizations actively seeking or who have gained certification

that they were separately assessed by a third party within accredited scope under ISO 9001. The impact of this is that companies are only able to enter the Defence Contractors List if their AQAP registration has not yet expired or if they have 3rd party certification covering the product/service to be supplied.

Third party certification

The number of 2nd party assessments should be reduced as a consequence of third (3rd) party certification. As long as an organization's quality management system is being independently judged as complying with the relevant quality standard, customers of that organization should have sufficient confidence in them so that 2nd party assessment is not required.

Current assessments in Europe include those contained in Table 15.3 (based on Marshall, 1994).

15.4.3 Benefits of certification to BS EN ISO 9000

Benefits include:

1 A continuous evaluation by external quality professionals, who objectively audit your quality management system in order to ensure that it functions as you want it.
2 A marketing edge by advertising and promoting demonstrated commitment to quality standards.
3 The provision of the basis for continuous quality improvement, as it ensures that quality-related data are developed and made available in order to manage quality more effectively.

General audit process for certification

Before the audit process can begin, the organization submits a questionnaire and application form, together with the application fee. Also before an audit is arranged, the organization submits to the certification body a quality manual, which is deemed to comply with both the relevant parts of the standard, and quality guidance material (if any) for the industry or commerce sector concerned. A detailed appraisal is made of the applicant's documentation and any significant deviations or omissions are noted in a report which is sent to the applicant. This enables the applicant to amend the documentation before the quality audit visit.

A planning visit is arranged and the audit team leader discusses with the applicant the costs and resources required for the audit, together with target timescales. Each audit is unique and occasionally more than one visit is required. When the applicant's quality manual has been approved, the formal audit is arranged, which involves an in-depth appraisal of the organization's procedures for compliance with the appropriate part of the selected standard — usually BS EN ISO 9000 — and the relevant guidance material. The organization is expected to demonstrate the practical application of its documented procedures.

Some certification bodies carry out pre-assessment visits to ensure that management and staff know the purpose and scope of the audit. The formal audit covers every aspect of the organization's quality system, although the determined scope may limit the actual coverage in everyday operations of the organization. Audit procedures are normally adhered to those required by the certification body.

There are three possible outcomes of a formal audit:

1. Immediate certification — as when all requirements of the adopted quality management standard are met.
2. Certification subject to discrepancy — as when a minor error has been detected.
3. Non-certification — as when there are too many significant errors indicating that the developed procedures are insufficiently demonstrated to meet the requirements of the adopted quality management standard.

The initial audit is followed by regular planned surveillance visits each year, or perhaps biennially (as the organization determines necessary) and for as long as the certification is required. Some certification bodies require a complete formal audit every three years and this should be encouraged.

Examples of organizations that are empowered to certificate as independent assessors include BSI-QA, Yarsley QA, etc.

15.4.4 *Accreditation*

Accreditation is the mechanism for ensuring the standard of third party certification bodies. This discussion will focus mainly on British developments. The July 1982 White Paper cmd 8621 (UK) suggested that a national accreditation scheme was necessary and that a new system should have a recognizable accreditation mark. It was suggested that there was a need for independent certification schemes in addition to BSI-QA. Accreditation was to be the mechanism for regulating these independent certification schemes.

The DTI Register states that 'Accreditation is a status awarded to certification bodies by the Secretary of State. He acts on the advice of the NACCB'. This was set out in the 1982 White Paper, *Standards, Quality and International Competitiveness (UK)*.

NACCB — National Accreditation Council for Certification Bodies

Following discussions between the DTI and the British Standards Institution, a memorandum of understanding was published in May 1984. This resulted in the set up of NACCB. Although it was set up as a council of BSI, the NACCB is totally independent of that part of BSI that undertakes QA audit and certification (BSI-QA) and became operational in 1985.

Britain first produced the equivalent of what is now the EN 45000 series of standards, which cover the operations of the NACCB. It is this standard which the NACCB uses to validate the certification methods of certification bodies. Other European countries have set their equivalents to the NACCB developments. The closest comparison with NACCB is Raad voor Certificatie of The Netherlands. Problems of accreditation have been highlighted where the NACCB will not accept applications from non-UK certification bodies. However, The Netherlands body has accredited BSI-QA under certain scopes.

There are four categories of certificate eligible for accreditation:

C1   Certificate of Quality Management Systems to BS EN ISO 9000 series.
C2   Product Conformity Certificate, e.g. Pascal.
C3   Product Approval.
C4   Certification of Personnel Engaged in Quality Verification.

So far (in the UK at least) certification bodies have only been accredited under C1 and C2. Both 2nd and 3rd party certification bodies are eligible for accreditation by NACCB. To date, no 2nd party certification bodies have been approved and perhaps never will.

Certification — scope

In certification, the scope refers to the coverage that the quality system applies to, i.e. the product line, department or specific physical entity, that can be isolated by a real or pseudo-boundary. This means that the assessment will have a limited boundary, as determined by the organization. The scope could also mean the entire operations of the organization — which should be classed as normal — and can include manufacturing or service entities. Problems arise here, because even when the quality system has been certificated, the actual scope may not be made apparent in any advertising supporting the certification, i.e. the quality mark, as used on any product produced by the organization.

An advantage of the certification to scope is that the organization can pilot a certificated quality system and develop a piecemeal approach to include other parts of the organization at some future time. It can also provide a controlled and meaningful way to focus the quality system and develop reinforcement of good quality management practices.

All organizations that have been certificated will be contained within the DTI register designed for the purpose. Their scope will also be published in that register.

In the United Kingdom, the NACCB has accredited various certification bodies, e.g. BSI-QA. As part of this accreditation, the scope of the certification body is determined — areas of work or activities that the accreditation body has approved a given certification body to carry out.

The DTI Register publishes each year all organizations that are currently certificated by all certification bodies which have had their quality management system successfully assessed. The Register does not differentiate between organizations that were certified by certification bodies inside or outside of scope. It is used by customers to determine who is capable of providing consistent products/services that will meet their needs, relevant to their commercial sector.

Where a certification body certificates out of scope, the NACCB has seen this problem and has accepted that suppliers certified outside scope can use the Tick mark and gain publication in the DTI Register, when the certification body has successfully increased its scope to include that organization's certification. The implications are that scope, or lack of it, could cause at least embarrassment and at most loss of contracts.

Accreditation — scope of a certificating body

An important issue of accreditation is the scope of the certificated body. The accredited scope defines the areas of work that the accreditation body has approved a particular certification body to carry out, e.g. to date, no certification body had been accredited by the NACCB to certify companies for either education and training or software. Any certification body can issue a certificate in compliance with the BS EN ISO 9000 series. If the certification body is operating within accredited scope, then the certificate is allowed to show the NACCB 'Tick' of approval, as well as the certification body's logo.

A certificate without the NACCB support is deemed to be of far less value. Customers can have no confidence in the methods and staff employed in the certificate's scope. This does not stop certification bodies issuing certificates to companies outside their accredited scope and unfortunately accounts for 70 per cent of the certificates issued.

Prior to the 1993 edition of the DTI Register, an inspection revealed that the majority of certificates were issued outside the accredited scope of the certificating body. Publishing certificated organizations within certificated scope only would result in a much reduced DTI Register. However, publishing only within certificated scope would encourage certification bodies to improve their operations and extend their accredited scopes appropriately. Many certificated bodies have very restricted scopes. We can expect the number of accredited bodies and their scopes to increase over time. This should ensure consistent standards of certification — at least within the accredited scope of the bodies concerned.

The process of accreditation

The criteria to be met by certification bodies seeking accreditation to certificate quality systems are set out in EN 45012. The DTI Register suggests that the

> main requirements for accreditation, as for inclusion in the Register, are that the certification body should have a representative independent board . . ., should possess quality systems to BS EN ISO 9000 or equivalent and should have staff competent for the work . . . and requirements as to documentation and regular surveillance of certified companies.

The requirements for an organization to gain accreditation are set out in Appendix 3 at the end of this chapter.

Basic criteria for all certification bodies applying for accreditation can be seen in the DTI Register. However, revised or additional requirements may be specified by the Accreditation Council as directed by the Secretary of State.

The accreditation process for an intending certification body must follow a similar type of process as certification, in order to gain accreditation. Before the audit process can begin, the organization submits a questionnaire and application form, together with the application fee, in much the same way that an organization would submit an application to a certification body. Also, before an audit is arranged, the intending certification body presents a quality manual, which is deemed to comply with both the relevant parts of the standard (EN 45012) — as discussed earlier.

A detailed appraisal is made of the applicant's documentation and any significant deviations or omissions are noted in a report which is sent to the applicant. This enables the applicant to amend the documentation before the quality audit visit.

A planning visit is arranged; the audit team leader discusses with the applicant the costs and resources required for the audit, together with target timescales. Each audit is unique and occasionally more than one visit is required. When the applicant's quality manual has been approved, the formal audit is arranged, which involves an in-depth appraisal of the intending certification body's procedures, for compliance with the appropriate part of the standard. The organization is expected to demonstrate the practical application of its documented procedures. Also, the NACCB would audit the process used by the intending certification body. This includes an analysis of demonstrated competence by the staff of the organization while auditing a supplier's quality management system against a standard, such as BS EN ISO 9000. Usually about five such audits are carried out. The areas covered by these audits would determine the scope of the accreditation given to the organization. The formal audit covers every aspect of the organization's quality system.

There are three possible outcomes of a formal audit:

1. Immediate registration
2. Registration subject to discrepancy
3. Non-registration

The initial audit is followed by regular planned surveillance visits each year for as long as the accreditation is required.

15.5 Auditing to BS EN ISO 9000

ISO 8402 states that a quality audit is a 'systematic, independent examination and evaluation to determine whether quality activities and results comply with planned arrangements and whether these arrangements are implemented effectively and are suitable for achieving objectives'. A quality audit's main purpose is to evaluate conformance to the adopted standard. It is not an audit of the quality of output, but an audit of the processes that constitute the audit scope. As an outcome of this, the basis for conformance — as well as non-conformance — should be evaluated in order for continuous improvement to occur. Auditing, therefore, should not be seen as *keeping* the status quo, although many organizations use it this way.

In order for the auditing to have some credibility, it is necessary for an independent person to carry out the audit. This means that the audit must be carried by any trained individual who is not the owner of the process or is not the process owner's supervisor.

Auditing can take the form of:

1. *Self-assessment* — for the purpose of evaluating one's own processes. Self-auditing is most useful when conducted to assist the independent audits carried out in response to the requirements of a given standard. Internal testing provides staff with controlled experiences that can only enhance their independent audits carried out elsewhere in the organization.
2. *Independent assessment — internally*. This is as (1) above, but generally in response to the requirements of a given standard, and would be carried out in another part of the organization.
3. *Independent assessment — externally* — of a 3rd party. This used to occur in the situation of 2nd party assessments. As indicated elsewhere, 2nd party assessments are giving way to 3rd party practices.

People need to be trained in auditing processes, as they are the most widely used activity in maintaining the quality management system. Interpretation skills — of the standard and what is actually carried out in the field — are developed as a consequence of the generated experiences. It is these skills that will ensure that the system is tested appropriately, so that non-conformities are discovered promptly and effectively.

Auditing is part of the management of change, where quality records provide past data, and auditing the discovery and testing process. The outcome of this provides corrective action requirements and opportunities. This also provides a powerful basis for the management review mechanism.

There are various reasons that underpin the purpose of a quality audit. These include:

1. Systematic tracking conformance to stated specifications as regards the adopted standard.
2. Determining inadequacies in the management of the process under examination.
3. Determining procedural inadequacies.
4. Identifying opportunities for improving procedures and processes.
5. To gain *real* data about the processes under scrutiny that are uncompromisingly accurate, so that the conclusions that rest on them have the power of objectivity and can therefore be relied upon.

Procedures for conducting an audit are required by BS EN ISO 9000.

There is a need to document planning and the operation of the audit. This includes:

1. The development of a plan that encompasses the full scope of the audit schedule.
2. The choice of personnel to control and conduct the audit.
3. Documenting and recording observations.
4. Determining, implementing and confirming corrective actions.
5. Reporting the planning, conduct and outcomes of the audit.

Reporting will need to ensure that it is clear and concise.

The individual responsible for a particular audited area will generally not be a quality professional. Therefore, the language used and the technical nature and level must reflect this.

A word on corrective action. Corrective actions will generally be carried out by individuals who are audited, rather than the auditor. Consequently, auditing is somewhat similar to a policing action. But if the organization is fully committed to quality, then the motivation to correct any non-conformities is likely to be very high, irrespective of the obligation placed on them by the BS EN ISO 9000 standard.

The human relations aspect of auditing creates difficult problems to overcome. For example, in some organizations auditing is seen as a means to apportion blame. This cannot be tolerated in a quality-oriented organization. The blame culture — where resentment can inhibit information flow — needs to be addressed, since many of the decisions in auditing are based on the opinion of an outsider. A major requirement is the development of positive attitudes of the auditors, the audited and of the management of the audited areas. This can derive from the positive reinforcement that auditing can offer — over a period of time. Advantages of auditing include:

1. Familiarization of internal processes by individuals who may not have been involved in the development of that process — provides some objectivity.
2. Appreciation of the complexity and technical nature of processes in the organization.
3. Training and educating the workforce to become responsible for the development and maintenance of efficient work processes and practices.
4. Increased team building through greater cross-boundary communication.

15.6 BS 7850 — total quality management

This standard was developed by the Quality, Management and Statistics Standards Policy Committee of BSI. BS 7850 is divided into two parts:

Part 1 — Guide to management principles.
Part 2 — Guide to quality improvement methods.

Part 1 is the focus for this discussion. This standard gives guidance to 'management on ways to make the organisation structure, management system and quality system more effective in meeting organisational objecties'. One of its main achievements is that the standard actually declares a definition for total quality management — 'the management philosophy and organisation practices that aim to harness the human and material resources of an organisation in the most effective way to achieve the objectives of the organisation' — which seems to be a rather wide definition.

The application of the standard requires consideration of:

1. *Policy and strategy* — This includes the organization mission, leadership and commitment and the development of directing objectives.
2. *Management* — This includes organizational structure, the management system, information system and communications.
3. *Improvement* — This includes working environment, measurement, improvement objectives and monitoring and review.

Little experience has been reported to date. This is rather disappointing, as the standard has so much promise.

15.7 Other standards that may be useful to consult

15.7.1 *BS 7750 — environmental management standard (ISO 14001 in draft — expected 1996)*

This standard was first published by BSI in 1992. Its production indicated the growing pressures and concerns for managing the environmental effects of products and services both domestically and internationally. It provides a means to focus quality management, health and safety aspects and concern for the environment in a more cohesive way by addressing the development of a framework that is proactive and forward-looking. One possible shortfall is that BS 7750 does not require an environmental policy statement. BS 7750 is concerned with environmental issues resulting out of the consideration and implementation of the design, manufacturing process, use and eventual disposal of the product. In this regard, it has implications for the disposal of packaging as well as the product itself. Concern is also demonstrated for the workforce, as it also incorporates procedures to deal with health and safety and accident prevention.

The standard requires three manuals to be developed:

1. Environmental Management Manual — Similar to the BS EN ISO 9000 QMS manual.
2. Register of Environmental Regulations — Affecting the organization in the pursuit of its business goals. The Register develops the policy and appropriate environmental regulatory requirements facing the organization.
3. Register of Environmental Effects — Develops the exact issues perceived to be facing the organization and indicates how the organization addresses and attempts to control these issues.

As with the BS EN ISO 9000 standard, a manager must be appointed who will be responsible for the development and maintenance of the manual and to ensure that it is up to date. In many cases, this appointed manager will be the same person as the individual responsible for BS EN ISO 9000. However, in large process industries, the demand for a separate environmental manager would be likely.

In July 1994, the British NACCB invited certification bodies to apply to them for accreditation. This was a major step forward, as until the standard could be certificated against, little recognition could be given to organizations that met the standard's criteria. This resulted in eight certification bodies receiving their certificates of accreditation on 8 March 1995.

Again, the process of accreditation would mean that a certification body would need to develop a track record of auditing experience. Also, the information and process of evaluation had not been finalized and the experiences would be taken into account to develop an effective assessment process. Auditing to the standard is like auditing to BS EN ISO 9000, but BS 7750 does not indicate how often auditing needs to take place. It is suggested that audits will be required to be carried out at intervals between one and three years (much similar to the Eco-Management and Audit Scheme (EMAS)) depending upon the strength of the perceived environmental impact, e.g. every year for a site with a high perceived impact, and every three years for a site with a low impact, although this is only a guide. How this assessment is judged depends on the circumstances.

Benefits of implementing BS 7750 include:

1. Reduction in occurrences and the cost consequences of waste, accident and/or error leading to environmental problems.
2. Increase in cost savings due to recycling.
3. More effectively designed product that results in longer product life, and hence savings on purchasers and raw material usage.
4. External recognition of being *environmentally friendly*, which provides a further marketing edge.

15.7.2 *Investors in People (IIP)*

This is not a standard, such as BS EN ISO 9000. Nevertheless, its importance is increasing in use and from its quality philosophy. Keen (1993) indicates that IIP offers 'a national standard, based upon extensive research into best employer practice in developing all people to achieve quality goals'. It is therefore more related to TQM than to BS EN ISO 9000, because it addresses human, rather than technical aspects. It was first developed in 1990 and used the best practices of successful organizations in the United Kingdom. As at October, 1994 some 10,000 organizations in the United Kingdom had become certificated or are working towards the standard (Chapman, 1994).

IIP has four key principles:

1. Making a commitment to developing people.
2. Reviewing the training, development and education needs of all staff concerned with quality.
3. Taking relevant action to meet those determined needs.
4. Evaluating outcomes in order to ensure that staff are developed effectively and that this development helps the organization secure its business goals.

These provide benefits that are synonymous with TQM, and provide an increased business performance, coupled with increased communication and the development of an innovative and learning culture. This tends to lead to lower waste and lowered cost. In this regard, IIP and TQM are seen to be working towards the same end.

The processes of implementation and certification include:

1. Understanding and anticipating the expected organizational impact the standard will necessarily have.
2. Generating commitment, especially from top management.

3   Evaluating the standard against present practice and determining the reasons for the gaps in performance.
4   As an outcome of (3) above developing an action plan to detail resources, timing and people involvement.
5   Working towards successfully implementing the action plan.
6   On-site assessment by an independent and qualified assessor.
7   Recognition as an IIP organization.

IIP can be seen as a recognition of the people culture towards quality. The mechanisms of assessment are similar to that found in BS EN ISO 9000. Any organization that has achieved certification to this standard can do no harm in seeking to certificate to the IIP standard.

15.7.3 *European Quality Award (EQA)*

This award was first launched in 1991. It seems to have been launched in response to the successful equivalents found in the United States — the Malcolm Baldridge Award — and in Japan — the Deming Prize. All three awards are firmly focused on rewarding successful TQM efforts.

The EQA is broken into the elements found in Table 15.4. What is very interesting is that the award gives 41 per cent to external performance elements in terms of results, the customer and environmental considerations. A self-assessment exercise provides a means by which these elements can be evaluated and reviewed systematically. This award takes quality standards further, for example, BS EN ISO 9000 could be used to ensure that the organizational policy and processes are as effective as possible.

15.8 Implications of the application of BS EN ISO 9000 — a service approach — higher education

Recently, some interest has been developed by individuals who see the positive benefits of implementing BS EN ISO 9000 in an educational institution as outweighing the potential drawbacks. On this point, the secretary of the NACCB confirmed that there had been a sudden upsurge in interest from institutions about BS EN ISO 9000, but few agreed upon 'what the standard meant' (Times, August, 1991). Others see BS EN ISO 9000 implementation as a way by which further bureaucratic control can be increased by upper management to encroach on the lecturer power base. James Tannock, of the University of Bristol suggested that 'the move towards BS EN ISO 9000 was a nasty and insidious attempt to impose bureaucratic standards derived from industry on academic departments' (Times, July, 1991). Others have developed their own interpretation and successfully applied the standard, e.g. University of Wolverhampton. Could these resistances be a reflection of the symptom, rather than a reflection of the true problems

Table 15.4 Breakdown of EQA standard

Element	Percentage breakdown
Leadership	10
People management	9
Policy and strategy	8
Resources	9
Processes	14
People satisfaction	9
Customer satisfaction	20
Impact on society	6
Business results	15

espoused by BS EN ISO 9000 implementation? It does serve to illustrate, however, that far from providing a panacea to solve some educational institutional ills, it can in fact develop some new ones.

Tuckman (Times, August, 1991) in a letter in reply to a July 12th article on quality, stated that 'While it is welcome that the NACCB question the relevancy of standards developed for the manufacturing industry being applied to universities, polytechnics, and colleges it is unlikely that from their vantage, they will pose fundamental objection'.

Tannock (Times, August, 1991), on the other hand, would rather see educational institutions carefully consider the introduction of the philosophy and practices of TQM, which 'shares the ethos of excellence which underlies higher education'. Buckingham (Times, November, 1991) suggests that there is 'confusion and consternation' in the way the translation of BS EN ISO 9000 is applied to educational institutions and refers to BS EN ISO 9000 as a 'straitjacket'.

Oakland (Times, August, 1991) was 'interested by the somewhat negative reactions of the "experts", quoted in The Times HES, July 12, 1991, to the introduction of TQM and BS EN ISO 9000 to higher education'. Oakland further suggests that he has not met one organization that 'cannot use the BS EN ISO 9000 quality system' and wryly states that there is no magic about education. If we believe Oakland, we can fully apply the BS EN ISO 9000 standard to educational institutions. So why the reticence featured in the press and elsewhere? Rooney (Times, August, 1991) was also puzzled by this. She suggested that BS EN ISO 9000 does not impose bureaucratic standards and it is the interpretation that creates the bureaucratization level. Heavily bureaucratized organizations therefore would interpret the standard in a similar way.

Rooney goes on by saying that it is 'vitally important that institutions wanting to implement BS EN ISO 9000 have suitable guidelines to help them do so and ... that the potential third party assessors of BS EN ISO 9000 systems for education are in fact competent to assess such systems properly'. This brings out the point that the quality 'professionals' within the educational environment need to know how to interpret BS EN ISO 9000 effectively.

The BS EN ISO 9000 standard has been much criticized by its inflexibility on dealing with its application to service industries and more particularly to educational institutions (Saunders and Walker, 1991). Much of what has been published about the implementation of BS EN ISO 9000 has so far been derived, primarily from the manufacturing industry, where standards of one sort or another have provided the framework for the 'mechanical' nature of their working environment.

Although Saunders and Walker (1991) discuss the application of TQM to tertiary education, we can find some justification to accept that the implementation of BS EN ISO 9000 can have similar impacts. The basis of the language of BS EN ISO 9000 originated from a manufacturing environment. Here, terms such as product, manufacturing, non-conforming product, test, installation, etc., are unfamiliar and do not correspond with the accepted nomenclature of an educational institution.

The language, being the fundamental means of communication, would provide the basic 'challenge to translate these concepts into terms that will be familiar to education' (Saunders and Walker, 1991). The use of a machine analogy provides a means for understanding the problems of interpretation and therefore meanings attached to various terminology contained within the standard. Unless the meanings attached to these descriptive terms are determined, then worthwhile replacements, in educational terms, cannot be realized (see further Sawada and Calley, 1985).

Since we have now made some comparisons between business and educational institutions and the reservations some individuals have about the transference of BS EN ISO 9000 to education, let us discuss the implementation of BS EN ISO 9000 purely from an educational point of view. The major differences that seem to exist between business and education are that many business (production) organizations can be referred to as mechanical bureaucracies, whereas educational institutions can be viewed as professional bureaucracies. In this perhaps lies the fundamental problems that surround the potential implementation of BS EN ISO 9000 in educational institutions. Proof that the standard works and can be interpreted effectively is going to be paramount for its wide acceptance in education circles.

Educational institutions, by their very nature, have a far wider standardization platform than their manufacturing counterparts. It is just this standardization experience that should allow quicker assimilation of BS EN ISO 9000, when a suitable guideline has been developed and adopted. The identification of the many problems associated with the implementation of BS EN ISO 9000, including the interpretation of the standard itself, suggests that the standard can be a vehicle for continuous improvement and only part of a larger problem of quality cultural change.

The application of the BS EN ISO 9000 standard to educational institutions is seen as positive by some individuals and negative by others. Individuals who oppose the implementation see it as bureaucratic and doing very little to increase the quality of the products and/or services produced.

It has been argued that since the education system has been involved in developing programmes of learning by relating these to given 'quality' standards and criteria, that they are and have in fact demonstrated their commitment to quality. So why introduce BS EN ISO 9000, when the quality exists already? Those individuals who are pro implementation of BS EN ISO 9000 can understand the reticence generated by other people, but determine that the eye of quality exists with the customer and not with the so-called expert providers.

Could the implementation of BS EN ISO 9000 generate resistance because of fear, for say lecturing staff, by being made more accountable; and will this try to ensure that they are more effective in their academic pursuits? At present the amount of data and limited discussion does not indicate which argument to believe. Where BS EN ISO 9000 has been implemented, there is a suggestion that the staff readily accepted the quality cultural change. Since actual implementation has till now been rather isolated, it would seem that it is too early to make judgements or a meaningful analysis.

BS EN ISO 9000 is here to stay, and it will increase in its use in the educational sector. The challenge is to make it work **for** education not **against** it.

15.9 Chapter review

Quality management standards are seen as a major pillar supporting the drive for continuous quality improvement through TQM. The quality system is designed to provide both the support and mechanism for the effective conduct of quality-related activities in an organization. It is a systematic means to manage quality.

BS EN ISO 9000 is seen as a voluntary management standard. BS EN ISO 9000 is not a standard to manage products *per se* as it is management-process-oriented. This means that you cannot use products in themselves as a basis for the demonstration of adherence to the BS EN ISO 9000 standard. However, increasingly, organizations certificated to the standard ensure that they are supplied by an organization which is also certificated to the standard. This creates an overwhelming pressure for suppliers — especially to government departments — to become certificated, whether they want to or not. The basis for the development of BS EN ISO 9000 can be squarely placed on its military background. The change came about in 1984, when the British Standards Institution managed to persuade the International Standards Organisation in Switzerland to develop a generic quality management standard for world-wide use, based on its revised standard of 1979.

BS EN ISO 9000 is broken into four parts. This provides a means for targeting the actual standard application, after consulting the guidelines in BS EN ISO 9004. These parts are:

1. 9001 — Model for quality assurance in design/development, production, installation and servicing.
2. 9002 — Model for quality assurance in production and installation.
3. 9003 — Model for quality assurance in final inspection and test.
4. 9004 — Quality management and quality system elements — guidelines.

The revision of BS EN ISO 9000 has come about after much debate and deliberation over the derived inadequacies of the original standard (1987). Arguments against service industry adoptions of the use of BS EN ISO 9000 were that it laid emphasis on the development of bureaucratic and paper-driven systems, rather than creating a culture that provided for the application of flexibility and innovation. What the updated standard attempts is the rationalization of the applied process, rather than the content of the requirements of the standard. The new standard tries to nullify this by providing an organization with the capability of choosing its own quality management system and subsequent requirements.

- *Certification* — This is where 1st, 2nd and 3rd parties evaluate a company's quality system against some specified or stated standard or manual.
- *Accreditation* — This is the mechanism for ensuring the standards of 3rd party certification bodies. The body's certification system is itself evaluated against a certification standard. One such standard is the EN 45011/12 standard.
- *Certification body* — An impartial body either governmental or non-governmental, possessing the necessary competence and reliability to operate a certification system, and in which the interests of all parties concerned with the functioning of the system are represented.

Accreditation is the mechanism for ensuring the standard of 3rd party certification bodies. This discussion will focus mainly on British developments.

A quality audit's main purpose is to evaluate conformance to the adopted standard. It is not an audit of the quality of output, but an audit of the processes that constitute the audit scope. As an outcome of this, the basis for conformance and non-conformance should be evaluated in order for continuous improvement to occur. Auditing, therefore, should not be seen as *keeping* the status quo, although many organizations use it this way. Auditing can take the form of:

1 Self-assessment.
2 Independent assessment — internally.
3 Independent assessment — externally — of a third party.

BS 7850 — Total Quality Management. This standard was developed by the Quality, Management and Statistics Standards Policy Committee of BSI. BS 7850 is divided into two parts:

Part 1 — Guide to management principles.
Part 2 — Guide to quality improvement methods.

This standard gives guidance to 'management on ways to make the organisation structure, management system and quality system more effective in meeting organisational objectives'.

BS 7750 is concerned with environmental issues resulting out of the consideration and implementation of the design, manufacturing process, use and eventual disposal of the product. In this regard, it has implications for the disposal of packaging as well as the product itself. Concern is also demonstrated for the workforce, as it also incorporates procedures to deal with health and safety, and accident prevention.

Investors in People (IIP) is not a standard, unlike BS EN ISO 9000. Nevertheless, its importance is increasing in use and from quality philosophy. It is more related to TQM than to BS EN ISO 9000, because it addresses human, rather than technical aspects.

The European Quality Award was first launched in 1991. It seems to have been launched in response to the successful equivalents found in the USA — the Malcolm Baldridge Award — and in Japan — the Deming Prize.

15.10 Chapter questions

1 Discuss the term quality management system.
2 You are a quality management consultant. Explain the basic requirements of BS EN ISO 9000 to a newly appointed manager.
3 There are no real benefits to implementing BS EN ISO 9000. Discuss.
4 Compare and contrast certification and accreditation.
5 Discuss the process of certification to BS EN ISO 9000.

6 Discuss the process of auditing a quality management system.
7 Choose one of the following and discuss its contents and implications of implementation:
   (a) BS 7750
   (b) BS 7850
   (c) Investor in People Standard
   (d) The European Quality Award

APPENDIX 1 — BS EN ISO 9000 Quality System Elements

The quality system elements

Management responsibility which is inclusive of:
- Quality policy
- Organization
- Responsibility and authority
- Verification resources and personnel
- Management representative
- Management review
- Quality system
- Contract review

Design control which is inclusive of:
- General requirements
- Design and development planning
- Activity assignment
- Organization and technical interfaces
- Design input
- Design output
- Design verification
- Design changes

Document control which is inclusive of:
- Document approval and issue
- Document changes/modifications

Purchasing which is inclusive of:
- General requirements
- Assessment of subcontractors
- Purchasing data

- Verification of purchased product
- Purchaser-supplied product

Product identification and traceability

Process control which is inclusive of:
- General requirements
- Special processes

Inspection and testing which is inclusive of:
- Receiving inspection and testing
- In process inspection and testing
- Final inspection and testing
- Inspection and test records
- Inspection, measuring and test equipment
- Inspection and test status

Control of non-conforming product which is inclusive of:
- Non-conformity review and disposition
- Corrective action

Handling, storage, packaging and delivery

Quality records

Internal quality audits

Training

Servicing

Statistical techniques

APPENDIX 2 — BS EN ISO 9000 Higher Education Survey Instrument

Would you please evaluate the following statements in regard to BS EN ISO 9000 — the Quality Management Standard.

> Please indicate how important you find each of the following statements by choosing a point on the scale ranging from 1 to 4.
>
> 4   The statement is **absolutely essential** for judging quality in higher education
>
> 3   The statement is **desirable but not crucial** for judging quality in higher education

2   The statement is of **little importance** for judging quality in higher education
1   The statement is of **no relevance** for judging quality in higher education

Statement   **Importance Rating**

1   Ensuring that the policy for quality is understood, implemented and maintained at all levels in the organization is   ____
2   Defining the responsibility, authority and interrelation of all personnel who manage, perform and verify work affecting quality is   ____
3   The identification of and resource of verification requirements is   ____
4   The appointment of an individual to manage the requirements of BS EN ISO 9000 so that they are implemented and maintained is   ____
5   Management ensures the continuing suitability and effectiveness of the quality system is   ____
6   Documents that constitute the quality system are employed to ensure that the service conforms to specified requirements is   ____
7   Controlling and verifying course/service design is   ____
8   Controlling documents and data that relate to the requirements of BS EN ISO 9000 is   ____
9   Identifying educational product (students, results, media) in a way that indicates the conformance or non-conformance with regard to inquiries, inspections or examinations performed is   ____
10 Ensuring that educational product (students, results, media) that does not conform to specified requirements is prevented from inadvertent use is ____
11 The detection and elimination of potential causes of non-conformance of educational product and prevent their re-occurrence is   ____
12 Prevention of damage or deterioration to educational product in handling, storage and delivery is   ____
13 The verification of quality activities that comply with planned arrangements and the determination of the effectiveness of the quality system is ____
14 The identification of training needs that ensure that personnel performing specific assigned tasks are qualified on the basis of appropriate education, training and/or experience is   ____
15 Ensuring that service delivery is performed and verified in a way that meets the specified requirements is   ____
16 The identification of adequate statistical techniques required for verifying the acceptability of process capability and educational product characteristics is   ____

APPENDIX 3 — EN 45012 Clauses

All clause numbers refer to EN 45012 and must be enclosed in a quality manual:

Clause 1 — Sets out criteria that must be satisfied for the use of bodies concerned with recognizing the competence of certification bodies, i.e. NACCB

Clause 2 — Scope — specified range of activities — of accreditation must be defined unde 2.1 (see NACCB guidelines) and include products, services activities and quality systems

Clause 3 — All suppliers must have access to the services of the certification body, which must not be unduly restricted by financial or other conditions or be discriminatory

Clause 4 — The certification body must be impartial, have an appropriate structure and permanent personnel dealing with the every day administrative activities

Clause 5 — The governing body (Clause 4) is responsible for performance of certification as defined in EN 45012, to formulate policies, oversee implementation and finances of the certification body and setting up appropriate committees — relative to scope

Clause 6 — The following must be made available — organizational chart, description of financial support, certification systems documentation, and legal status

Clause 7 — Personnel must be competent to carry out the functions of the certificating body and the standard determines criteria for this to be assessed

Clause 8 — Documentation and control procedures should be maintained for the control of all documentation relating to the certification system

Clause 9 — Suitable records must be maintained

Clause 10 — Certification, surveillance and documented procedures covering initial assessment and subsequent assessment of a supplier's quality system are required

Clause 11 — Adequate certification and surveillance facilities are required

Clause 12 — The certification body must have a quality manual

Clause 13 — Adequate arrangements must be developed to ensure confidentiality of the information obtained in the course of certification activities

Clause 14 — Publication of a list of certificated suppliers with an outline of the scope of the certification of each supplier is required

Clause 15 — Adequate procedures for consideration appeals against decisions must be developed

Clause 16 — Undertake audits and periodic reviews of their compliance with the criteria of standard EN 45012

Clause 17 — Exercise proper control of Quality System certificates

Clause 18 — Require certificated suppliers to keep a record of all complaints and remedial actions relative to the Quality System

Clause 19 — Documented procedures for withdrawal and cancellation of Quality System certificates need to be developed.

Part 4

Integrated Quality Management — The Future

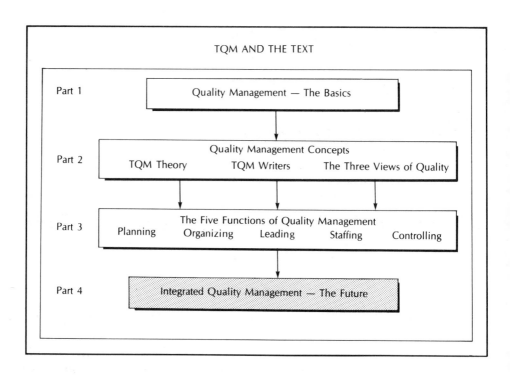

CHAPTER 16

Total quality management: Future issues that need to be addressed today

CHAPTER OUTLINE

- 16.1 Introduction
- 16.2 Integrated systems of quality management
  - 16.2.1 *Processes*
  - 16.2.2 *People*
  - 16.2.3 *Structures*
  - 16.2.4 *Technology*
  - 16.2.5 *Customers*
- 16.3 Ideological basis for future effective development and operation of TQM
- 16.4 Concluding remarks

16.1 Introduction

What will TQM look like in ten years? Will it still be here, operational, and possibly more important, will it be effective? This conclusion is a brief discussion about ideas and notions that may be seen to provide for the future developments and use of total quality management, and also to evaluate its potential in building an integrative quality management approach. Consequently, an understanding of some of the effects and implications of the cornerstones of TQM concepts may be useful in order to ascertain the future position and prominence of TQM. These issues include:

1 Integrated systems of quality management.
2 Processes.
3 People.

4   Structures.
5   Technology.
6   Customers.
7   Ideological basis for the future effective development and operation of TQM.

16.2 Integrated systems of quality management

Integrated quality management systems should, by default, include the use of BS EN ISO 9000, BS 7750, BS 7850 (UK denoted) and other respective quality management systems derived for enhancing the customer–supplier relationship. They will also include the integration of health and safety legislative requirements, such as BS 8750, the Health and Safety at Work Act 1974 (UK), and new regulations concerning European-wide legislation and, for example, BS 6143, the new process cost model. Together these standards — legal, health and work requirements, basic and advanced quality management techniques and systems — are seen to provide the future sphere of operation for quality-related activities in organizations.

Isolated but related standards and narrow management activities cannot be acceptable for the effective management of organizations in the future. TQM will no longer be seen as a singular quality management discipline with secular applications to organizations. It will be an integrated strategy involving independent disciplines of design, marketing, production, HRM, management and health and safety professionals internally and customers and vendors externally.

Integrated quality management means taking what is being developed today and applying it today, but in a more open, more responsive, more flexible, and more integrated way. It also means that the future of quality management undertakes to reduce the duplicate efforts made to satisfy one or more of these independent standards and to ensure the integrative development of quality management systems that can meet the ever-changing environments and quality requirements of both supplier and customer, internal and external.

The major differences between TQM today and what it will be like tomorrow can be seen in Figure 16.1.

16.2.1 *Processes*

1   *Time-based competition* — This is the overall time it takes a product to get from product conception to commercialization, and will be reduced dramatically using a process of standardized innovation. This is the result of collaborative internal processes, using concurrent and simultaneous engineering principles — in *all* disciplines and functions of the organization. Consequently, time to market is markedly reduced. Its power is derived from the effective use of new technology, mainly in the form of structured

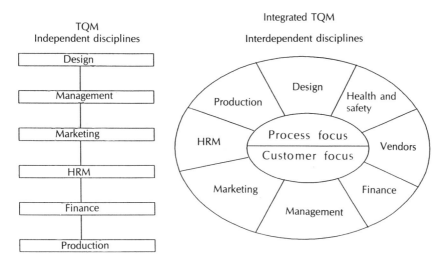

Figure 16.1 Integrated TQM.

information gathering, collation, analysis and storage, that provides an integrated and flexible capability.

2. *Competitive collaboration* — Through benchmarking will ensure market penetration in an increasingly smaller market. Pseudo-competitors, using integrated TQM, will see other organizations as knowledge machines that reduce the learning curve and increase the chance of survival. The paradox is that more information about the market will ensure that competitors target much more effectively, with lower direct competition, and increasing customer satisfaction. In this sense, integrated TQM means helping competitors to help you.

3. *The paperless quality management applications* — Will mean on-line auditing, process development and re-engineering, and the instantaneous application of SPC-related techniques. This is already a familiar sight in a number of forward-looking organizations today. What is significant is that concurrent development will be at such a pace that products will be developing as they are produced. That is, products will be changed through flexible manufacturing processes. In this, the notion of a single standard product will become meaningless, except that the notion of a single customer will become reality and there will still be the economies of scale derived as a consequence.

4. *Quality in services* — Will become the focus, as process management techniques becomes standardized. The use of engineering-type methodologies will be adopted throughout the organization in order to increase the effective throughput of product or information. In this way,

the engineering bias of these methodologies will become blurred and ownership and use of them will therefore increase. However, all customers will be expected to contribute to the effective management of the service and thus internal as well as external customers will have the responsibility to do so. What will be significantly different is the way customers are treated. They will therefore become equal partners in the management of their requirements, a case of customer value through involvement, not just as an input through data generation, but as ongoing, on-line developments.

5. *Interpretation* — There will be a move towards interpretation, rather than the focus on the collection of data, as the methods of collection become standardized in organizations. Here, flexible process design (similar to manufacturing design) and flexible product design will mean that standardized processes can quickly be developed and used to ensure customer satisfaction. It will also mean a greater need for the application of decision making at the point of need, and thus managers — especially middle managers — will be likely to take on a facilitating rather than an administrative role.

6. *Normalization of TQM practices* — The treatment given to integrated TQM practices will lead to it being viewed as the normative operational requirement. This contrasts with many of today's TQM applications that are seen as the application of a continuous series of 'testing' events that may result in a flexible and quality-oriented learning.

16.2.2 People

1. *Change agents* — Staff will become trained and educated to become change agents for themselves, rather than relying on quality management consultants or other professionals. The learning organization will be just that, learning about themselves and the way they operate. Data will be readily available through the use of high technology, and the skills, competencies, education and training will become more integrated as a cohesive feature, rather than ad hoc training measures or measures designed to provide what management expect staff to want or must want to know. Here, on-line computer-based learning systems that attempt to integrate theory and practice at the time of need will become the norm. High technology will become a training and educating partner, limited only by the creativeness of the trainers and computer-based learning (CBL) technologists.

2. *Responsibility management* — This will mean developing an individual focus, which is organizationally supported, and always customer-directed. It means the acceptance of responsibilities for the effective management of the design, development and ultimate satisfaction of customers, while consulting customers and supporters. Responsibility management means

giving employees the right to take charge, but is always directed at the enhancement of the customer/supply interface.

3 *Networking* — Internally and externally (JIT, flexible manufacturing systems). Communications patterns will resemble collegiate models. That is, each individual will be seen as an expert that other individuals can tap into. They will be seen as experts in the sense of managing their tasks and their ability to help others manage concurrent or their up/down-line tasks.

4 *Innovation–entrepreneurship* — The need and requirement for innovation–entrepreneurship will become the norm. Flexible working practices will not be sustained unless staff are given the right, as well as the support, to develop these operational functions.

5 *Learning* — The act of learning to learn will become the norm. Since change will be an everyday occurrence, the continuation of the task of learning about the performance outcomes of business processes, people, technology by the people who actually make the decisions will be ongoing. Learning will be an activity that will not be constrained by barriers, demarcation lines or management levels.

6 *Rewards* — Flexible structures, an increasingly multi-skilled work environment, and increasing technology use will introduce flexible reward structures that are not only fixed on output performances. However, output performances are still very important; it is the total contribution that will be evaluated. Rewards are likely to be based on the available *store* or personal inventory of technical, and social skills, competencies and appropriate task responsibilities successfully undertaken, that the organization can use effectively in both the short and long term.

7 *Self-directed leadership* — Much of what is achieved in the integrated TQM environment will be carried out by self-directed leadership; based on trust, confidence and hands-on multi-task knowledge, this would provide a rich background for developing and enacting responsible task management. This, coupled with technology use, will mean that self-directed leadership may not be confined to actual work-based locations.

8 *Knowledge management* — Management as we know it today will become a shop-floor phenomenon and requirement. However, unlike the currency of top management which will still be money, the currency for the rest of the organization will be knowledge, and this knowledge will not just be isolated to internal group, department or organizational settings. The knowledge-based requirement will take on a three-level orientation — that is, three levels up, three levels each side of each process and three levels back. Staff will need to know the process requirements of three processes forward and back in the production track, the requirements of three levels of management (theirs, top management and supplier) and the ultimate

customer orientation and requirements. Thus, the complexity of operation will increase and the need to manage the developed knowledge effectively is of prime importance.

9 *People investment* — Human resources will be seen as an investment, rather than as a cost associated with production. Consequently, the managerial mentality will have changed to incorporate staff as partners, rather than staff as subordinates.

16.2.3 *Structures*

1 *Organizational and group structures* — There will be a need to reconcile the application of highly structured organizational forms that create the base for efficient information storage, analysis and retrieval with flexible, integrated group structures which will be used to apply analysis and the developed strategic intent. These two seemingly different and difficult requirements will be judged to be effective and become more integrated, as the methods of quality management become more routine, and the need for flexibility in decision-making practices becomes more crucial. Structure will therefore increase as high technology lends its standardization power to assist groups to become more adaptable.

2 *Flexibility* — Normalizing TQM practices, increased education and training of the workforce and the need to increase speed of delivery of products/services will mean elaborate yet programmed changes to the structure of groups, and thus the organization. This means going beyond matrix management and into structureless management, perceived to reflect derived opportunities concerning individual accomplishments through group management.

3 *Outplacement* — Integrated TQM will not involve the requirement to have all individuals come together at the same location to work. This will be especially so in the management of knowledge processes, such as planning, finance, accounting and much of the development of marketing activities. The driver behind this development will be the cost associated with traditional job placement. Consequently, investment in new technology that provides detached but on-line and thus real-time connection will become the norm. This will result in a reduced core of staff that remain on-site who are involved in directly producing the actual product/service. All other personnel will be off-site.

16.2.4 *Technology*

Technology use in the quality-oriented organization will increase. It will increase in the use of higher technologies to manage the design, production and delivery systems of the product or service offered. There will therefore be an increase in the amount and complexity of information that individuals will have access to, and the tools used to manipulate this information. The requirements for training and education will also increase in response to the heavier managerial workloads of workers.

In this scenario, high technology forms the backbone to the future development of an integrated TQM. Disciplines will not be isolated because of information barriers, and thus politics and power will be markedly reduced, and become more overt. Staff will become technologists in their own right and will apply many of the manual tools used today, in a more structured and automated fashion.

Technology will be used to structure tasks and information in such a way as to meet the flexible requirements of the users. It will provide an interface where complex data are translated into much simpler adaptations, without losing meaning or accuracy and possibly more importantly will be available instantaneously on-line. The results of events will be recorded, stored, analyzed and be available for interpretation and decision making much faster than today.

The motivators for this will be fourfold. First, the satisfaction of the need to know more about the operational effectiveness of organizational processes, tasks and outcomes much more quickly; the second will be as a result of competitive strategies and capabilities being altered as a consequence of their use of this technology; the third is the cost of such technologies will reduce dramatically; and the fourth the customer will be expecting to become more involved.

16.2.5 *Customers*

Customer input is sought to provide an on-line focus for continuous improvement. Customer opinions and changing needs and wants will be identified through the ongoing dialogue. This will result through feedback technologies that consumers use at home, at work, wherever they use the product or service. Since the technology will mean the realization and use of on-line, real-time information, then these can affect the production process by using flexible standardization (differentiated production systems) practices and systems. Customers will therefore directly affect all processes in an organization, rather than through interfaces such as marketing.

Customer-related issues include:

1. *Speed of response* — Much of the focus today is reducing delivery time of product/service. In a product environment, this can be implemented carefully and gradually. However, in the service environment, the speed of response prior to and when the order is taken can determine the customer's perception of the speed of response of the whole product/service delivery. What will be important will not only be to increase the speed of

response due to these routine requirements, but also for non-routine ones. For example, customers are not restricted by internal rules and regulations and consequently perceive poor service if there is a delay in responding to their non-routine requests, even if they are seen as unreasonable by the individual staff member. Integrated TQM would ensure that the staff member is empowered to ensure the immediate satisfaction of the customer in any way possible. This can mean not adhering to normal and routine procedures.

2. *Extending service choice* — Increased competitiveness will mean giving customers increased choice. This means choice of how they are dealt with internally, flexibility in how and when they are serviced and their stage of service, etc. Gone will be the days when an irate customer would have to repeat their problem to different staff members, whilst being bounced around different departments, who offset their responsibilities in a 'it is someone else's problem' culture.

3. *Added customer service value* — Customer orientation will mean giving customers enhanced service attributes that may not be totally related to the service contract. Going beyond what is expected and delighting their customer will mean just this, adding service value, sometimes without the direct return of customer support in terms of purchase and loyalty.

16.3 Ideological basis for future effective development and operation of TQM

In order to obtain the most potential from the TQM philosophy in the future, the following should be developed as one singular integrated system:

- *Total Transformational Management* — The management of the quality culture.

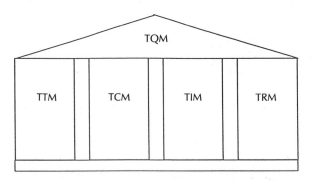

Figure 16.2 The new pillars of TQM.

- *Total Customer Management* — The management of the customer (internal and external).
- *Total Information Management* — The management based on on-line and up to date data.
- *Total Resource Management* — The management of all resources (physical, financial and human) in order to meet the previous three.

These elements that form this basis can be seen in Figure 16.2.

16.4 Concluding remarks

Total quality management is seen as a complex mix of ideology and resultant methods that reflects similar connotations found in the environment. TQM provides the basis for the effective application and management of methodologies to solve many of today's business-related problems. It also provides the basis for grasping opportunities, enhancing organizational strengths and reducing the effects of weaknesses and threats.

TQM is not a panacea for the ills of today's business. It can be seen as a structured, yet flexible approach; cohesive, yet a loose amalgamation of methodologies and applications; constructive, yet it breaks down barriers; understandable, yet complex; changing, yet standardized; stable, yet dynamic; co-operative, yet independent. TQM is all these and more. The key to the future, and more importantly the key to business strength, is to grasp TQM wholeheartedly and re-engineer all the processes in an organization so that they are all worked as integrated units. TQM is as much a worker concern as a management concern. Managers and workers need to collaborate as partners in the new quality revolution as the speed and complexity of change — as a result of increasing demands by consumers — will continue to accelerate. The future of the TQM-oriented organization is going to be the development of a learning organization that not only learns continuously about its customers, but also its people and operational processes.

Part 5

Cases and Problems

INFOCOM International: Customer service

75 Alderton Way
Crawford
Surrey

14 February 1994

Manager
INFOCOM International
11–17 Bygone Gardens
Manchester

Re: Problems with 486/DX2/66 Computer System

Dear Sir

On the 2nd of November 1993 I purchased the above computer system.

I am aware that you understand my various problems with this computer system but I will indicate the development of these problems as below:

1. When I opened the computer packaging and first turned on the system, at start-up I received an *HDD controller failure error*. When opened, the electrical connection for the hard disk was not in place. This was duly refitted. The computer continued to give an HDD error at intermittent times — even to date. I contacted your shop and was told to telephone the help-line. This I did. Consequently, a technician arrived (call No. 714774) and carried out various low-level tests. This resulted in an exchange of both the hard disk and the hard disk controller. This still did not rectify the problem of the HDD controller error.

2. Another problem that had developed was that the mouse pointer when in the Windows environment would intermittently 'stick' in hourglass mode. This became very annoying and time consuming, as the actual place of the pointer was difficult to judge.

3. Another problem that had developed was that the screen appears to reduce in size after 20 minutes when in a DOS programme. Effectively, this means that when working in DOS, I have to increase the height of the screen when working in text (the screen otherwise covers only about 60 per cent vertically). Unfortunately, when the text is viewed in graphics, it reverts to the graphics screen — which is now about 40 per cent larger than the screen

(vertically). When the programme is exited, and is returned to Windows the screen has to be moved back to cover the height of the screen — from 140 per cent to 100 per cent. This is a very annoying problem. I have the appropriate drivers installed. When the computer (monitor) is cold, the system works as expected — with no change between DOS and Windows programmes as far as video response is concerned. However, after 20 minutes or so, the problems cited above become evident every time the system is switched on.

4   On the 3rd December 1993 I returned 1 Mbx8 and purchased 4 Mbx4 SIMMs. I had no difficulty with fitting these.

5   On the 17th of December 1993, I again called in a computer technician about the same problem — *HDD controller failure*. This had resulted from a loss of data on the hard disk. The result of this visit was that the speed of the multi-IO card and the speed of the HDD were 'turned lower to medium'. This has failed to rectify the highlighted problem.

6   A new mouse was purchased from you on the 31st December 1993 — as it was suggested by you that it would rectify the mouse pointer problem. This still did not solve this problem and the pointer continued to 'stick' intermittently.

7   On the 29th January I returned the VL video card (1 Mb) and purchased in exchange a VL Pro (2 Mb) and a further 128 Kb cache memory (now 256 Kb — suggested that this would solve the HDD controller failure problem). The problem of the screen has not subsided and is constant every time.

Total amount invested in the computer system:

System	£2550
Extra SIMMs	£300
Mouse	£15
VL Pro video card	£300
128 Kb cache	£100
Total	£3265

With regard to the above computer system and its operating condition:

1   The system at present is unreliable.
2   There are faults that were detected right from the outset which have not been resolved.
3   Although various agencies have examined the equipment, they have not solved the continuing problems.

I have invested a lot of money and time and I have now come to the end of possible solutions. I would like you to replace the total system with a reliable one. If this is not possible, I would like to return the system and get my money refunded.

Yours faithfully

S R Stephens

Activity brief

1   You are the retail shop manager. What would be your response to the customer's request?
2   What immediate strategies would you adopt in order to ensure that the company met its responsibilities towards the customer?
3   What changes would you implement in order to prevent any subsequent failure occurrence?

ALPHA-ONE: Changing the quality-oriented culture of an organization

This case study illustrates the opportunities, problems and processes of culture change in an organization. Its focus lies with the developments surrounding cultural change in an already dynamic and customer-oriented organization. It indicates the time needed for culture change and the responsibilities of top management in this process. The case study is useful for the analysis of organizational change programmes and quality improvement schemes, and their implications for both management and other staff.

Introduction

ALPHA-ONE sees *culture* as indicative of the 'way we do things'. Consequently, the emphasis of the management of change has revolved around effective organizational cultural change. ALPHA-ONE is deemed to be one of the largest designers, manufacturers and marketers of computer-based information systems and related products and services in Europe. ALPHA-ONE is multi-located within the main business centres and has a staff of over 150, including computer technicians and other support staff.

This case study seeks to illustrate the cause—catalyst, design, process and results of an organizational development programme that was attempted.

Background

The following aspects provide the initiation or catalyst for the intended cultural transformation:

1. A merger, completed in November 1992.
2. Changing management philosophies and their ideologies.
3. Long-term goals.
4. The effect of those goals on the people in the organization.
5. The existing culture and environment.
6. The changing market requirements, both national and international.

Other development aspects include:

1   INPLAN acquired July 1994.
2   SawTech acquired December 1994.

The merger in 1992 provided management with an opportunity to do things differently. It was determined by ALPHA-ONE that a planned organizational development programme was the best way to achieve this. The basis and director for this change programme was the development of the ALPHA-ONE mission statement, in January 1992.

The ALPHA-ONE mission required a change to a more planned approach, which inevitably demanded 'clear goals, strategies, policies and objectives' to be percolated from a company-wide perspective to the individual level to ensure organizational commitment. The aim of the programme was to develop individual behaviour that reflected their contribution to their jobs around a framework of skills and knowledge. Outside market influences (customer requirements) also exacted a pressure on the change programme. It is the market forces that have required a change to customer service teams in order to act as integrated units to satisfy one of the published requirements of the ALPHA-ONE mission statement. To generate suitable customer servicing teams, a more flexible and integrated structure was needed and developed. The change programme was therefore both internally and externally oriented and prescribed an exact change that satisfied management and other staff and, more importantly, the customer base. The change programme necessitated the commitment of top management, in order for the change programme to gain credibility and enable managers to develop budgets for and spend on appropriate training programmes to support the cultural change.

The situation

The change process

An individual experienced in developing organizational development programmes was contracted to develop the change programme relevant to the ALPHA-ONE mission statement and its underlying management philosophies. The ALPHA-ONE philosophy behind the change process approach was to ensure a:

> Balance of technical, economic and social consequences of change with adoption of a 'plan, do, review' approach to its implementation.

It was decided at a very early stage that the matching of personal and organizational goals was a must to ensure individual commitment. Consequently, training programmes were focused on delivering precisely this. Once individual commitment was achieved, individual and organizational performance objectives could be enhanced.

An underlying philosophy of ALPHA-ONE is that people are their best resource, and are not limited to top management or other experts. The change programme is a process used to tap this mental energy and ensure individual participation organization-wide. This was demonstrated sufficiently so that multi-level task groups were established to resolve a variety of organizational issues. This ensured that all levels of the workforce would be committed to the outcomes from these task groups.

Essentially, the ALPHA-ONE understanding was that resistance to the outcomes would be decreased when individuals were given a chance to participate in the development of the resolutions proposed by the task forces. This is further enhanced when individuals also participate in the implementation process.

Culture change

ALPHA-ONE carried out a company-wide survey and interview programme, developed with the consultant, in order to establish a baseline upon which the development of the change programme could rest, i.e. the present or old culture. The other main objective of the survey was to elicit, directly from individuals within the company, what type of cultural change should be planned for, i.e. the new culture. The results of this survey can be seen below:

	Old Culture	New Culture
1	Individual	Team
2	Systems orientation	People orientation
3	Product focus	Customer and quality focus
4	Volume and sales	Marketing through satisfying customers
5	Short-term focus	Long-term focus
6	Slowness of response	Flexible and timely response
7	Mechanistically orientated	Organically orientated

The above provide the specific outcomes required from the cultural change programme. It illustrates the prior- and post-cultural demands. Did the change in fact provide those outcomes, or are there some that are satisfied and others yet to be tackled? To qualify this, inquiry was made as to the problems encountered and the reasons why some changes were successful and others not.

Obviously, not all strategies can be accomplished in the short term. ALPHA-ONE appreciates this point and have long held that the change programme is long term, although some benefits could accrue in the short term. This was a requirement, since early positive feedback gave the impetus for individuals to continue in the change programme by believing in the eventual outcomes and benefits that they would bring.

Impact of the change programme

The change programme itself presented a wonderful opportunity for energized and committed individuals to develop themselves personally, whilst creating a successful business ethic and an organizational culture that rewards everyone for effective performance. The careful management of a change programme, although it has its benefits, also has its costs. ALPHA-ONE has experienced both. The impact can be illustrated as follows:

- POSITIVE
1. Attitude of senior management to initiating and supporting the change programme.
2. Company-wide participation.
3. Enhanced customer focus and responsiveness.
4. Creation of customer orientation that is both internal and external.
5. Ongoing organizational culture development.
6. Open communication channels throughout organization.
7. The development of self-management practices.

- NEGATIVE
1. Ineffective change management techniques used, resulting in multi-level resistance to change.
2. Disorientation caused by change, especially at middle management level.
3. Immediate and short-term business became of secondary importance.
4. Demonstrated lack of consistent understanding and personal support for the change programme among strategic groups; infighting and resource allocation became a major issue.
5. The initial internal change programme designers, change agent and developers had left the organization.
6. Scepticism among all levels of the organization about the effectiveness of the process and content of the change programme.
7. Scepticism among all levels of the organization about the need for the change programme.
8. Some critical individuals did not think that the change programme was realistic in what it was designed to achieve and consequently failed to get the necessary commitment to its objectives.

1993 and beyond — What actually occurred?

In-depth interviews with the staff of Southern Region, ALPHA-ONE, provided an illuminating understanding of exactly what was achieved or not achieved. This was based on knowledge and experience gained from the introduction and implementation of the change process through to the present.

However, early in the interviews, it became apparent that all the cultural changes expected by ALPHA-ONE had not in fact been totally realized. This may not be an unreasonable occurrence, but may relate to the high expectations placed upon the company by the old Managing Director — John Dunn.

Below are tabled the strategies that became evident in 1993 that were the directing strategies for the organization.

Strategies

1. Specific focus on quality improvement; embed total quality management; vision for constant process improvement.
2. Middle management focus of a training and development programme to create empowerment and commitment to the principles of the change programme.
3. Encourage risk taking.
4. Create a more fully integrated organization.
5. More clearly defined customer relationships internally and externally.

The results of the interviews suggest that:

1. Only strategies (1) and (2) above have been developed to any degree. The other strategies have received little or no attention. Strategies (1) and (2) have been developed as follows:
   (a) Management have been trained in TQM techniques, and these are used to track customer satisfaction.
   (b) Customer relationships are deemed extremely important to ALPHA-ONE, hence the development for TQM techniques. ALPHA-ONE supports the need to ensure the customer competitive edge by ensuring that the customer uses the most appropriate computer technology effectively. A technology plan is being developed to account for this.
2. ALPHA-ONE has again changed its mission statement (four times in three years).
3. Staff of ALPHA-ONE view the initiation of the cultural change as a learning experience only for John Dunn, the old Managing Director. There was distrust evident in the change process and therefore staff ownership failed to be generated. They were not therefore fully committed to the change objectives.
4. Staff believe that an external influence, the change agent, has unnecessarily attempted to change the ALPHA-ONE culture.
5. Tom Chance, the initiator of the change programme, has now left ALPHA-ONE, and the staff indicate that along with him went the impetus for change.
6. ALPHA-ONE has developed into a risk-management philosophy rather than a risk-taking culture.
7. Since the business environment is changing extremely rapidly, then flexibility, change and performance are expected by all ALPHA-ONE staff.

This is determined as only being achieved effectively, when the staff feel they are in control, not being controlled by someone else.
8   The culture of ALPHA-ONE is seen now as similar to that prior to the change initiation. Overseas controls inhibit totally flexible structures that were seen as a base platform from which the change programme rested.

Conclusion

Tom Chance of ALPHA-ONE has experimented over three years on a culture change programme, whose objectives were not owned by his staff. Consequently, the programme itself has failed, especially since he left. Some aspects have provided the staff with greater opportunities, customer relationships, quality focus, etc. Nonetheless, without the programme driver, staff ownership of the change programme objectives, staff belief, stable ALPHA-ONE identity, etc., the change programme has accomplished little except create confusion, create expectations that were unfulfilled, and generate an anti-change mechanism, clearly seen in the present cultural environment which is similar to prior culture change attempts.

ALPHA-ONE staff feel they have not altogether failed in the change attempt, but it does suggest that to succeed fully requires their support as well as top management drive and determination.

Activity brief

1   You have just taken the job as Chief Executive Officer, what would you do to lessen the impact of the *failed* change programme?
2   What immediate strategies would you adopt in order to ensure that the company met its responsibilities towards the customer?
3   How would you implement a future quality change programme, in light of the company's experience?

Lotteries Commission: Managing quality with data

Introduction

As you walk into the headquarters of the Lotteries Commission, you are confronted by an atmosphere that may not only be unique in business circles, but one which is definitely progressive. What this atmosphere reflects is the performance, commitment and drive that would shame other better run organizations. It is an image that projects the modern concept and ideas about entertainment and perfectly describes what entertainment means to its customers.

The company result provided a productivity per employee of £0.67m. This result compares with the average top 200 companies in the United Kingdom of just £22 515.00. This result equates to the Lotteries Commission being 29.8 times more productive than even high-performing conglomerates.

To highlight the differences between 'normal' business types and how they operate, think about this:

1. The lotteries section of the Commission starts every Monday morning with a clean slate, i.e. no money available to pay the staff or creditors. This section accounts for 65 per cent of the Commission's business revenue.
2. The staff are expected to perform every single day with renewed vigour and commitment.
3. Franchisees expect the Commission to provide ongoing backup services, so that they can perform effectively.
4. Total knowledge about the company's activities is gained literally, second by second, through the use of a vast network of computers.

The mission statement of the Commission is distributed at every conceivable strategic location in the building to provide reinforcement of what the company is trying to achieve. It states:

> The Lotteries Commission is in the business of lotteries and entertainment. Its corporate aim is to reach profit and revenue targets through providing service.

Among other aspects of the business, a five-year strategy, which is internally developed and reviewed each year, indicates the sort of commitment that is expected from top companies. These philosophies include:

1 Employ the most suitable people and reward above the average.
2 Commitment to the development of people and the system and processes to help them carry out their jobs.
3 Use all measurements of effectiveness, financial, quality marketing and others in order to determine the effectiveness of the delivery of service.
4 Consistent application of management and marketing strategy and philosophy in the long term.
5 The development of a co-operative team-based achievement of goals and strategy.

This therefore provides the basis for the continued improvements within the company. Consider this: in 1991 a profit of £8.7m was made, in 1992 a profit of £20.6m and in 1993 a profit of £40.3m, almost 'absolute' proof that continued improvements are part of the company culture.

The visibility of the application of these philosophies can readily be seen when you walk into any of the Commission's office floors. There are open-plan offices, which are bright and cheerful, people do not just walk around from desk to desk, they run. There is raw energy, real energy that is channelled one way only — to satisfy customers through everything they do.

Performance needs to be measured, and the Commission is perfectly placed to carry out this activity. The Commission has two mini-frame computers that are the heart of their information system. Two are required in case one breaks down. Incidently, any break in on-line capability is normally attributed to the telecommunications network, which is outside of the control of the Commission. Because the information is always up to date and accurate, it provides the support of one of Deming's philosophies 'What gets measured, gets managed'. It also ensures that performance criteria can be set realistically, and immediate feedback can be attained to enhance individual capability, both in the office and in the field.

The system also allows the franchisees to gain through positive feedback, which in turn generates further commitment towards the customer.

The culture of the organization indicates it is a problem-solving culture, where ideas are generated by individuals, supported by relevant and accurate data and where winners are satisfied customers. There is a will to win, by generating data from without, to use effectively within.

Every individual has the right to access corporate information. Stephens says, 'It's more important that they get any and all information that is available to me, because it is they who carry out the day-to-day decisions'. 'Why are people so motivated?' I asked. Stephens replied,

> Motivation is seen as operating both top-down and down-up with equal responsibility for all, relevant to their positions, both from an expert and

a general point of view. Customer orientation is generated from within. Each individual becomes motivated because we have shared vision and shared goals.

A word about marketing. Every franchisee is surveyed at least once per year, sometimes twice, by an independent marketing consultant. Every franchisee is visited every four weeks. This is in addition to the constant communication that continues every day by the telephone line.

Does the Commission take any notice of what the franchisee has to say? Since the link between the customer and the Commission is the franchisee, then notice is certainly given to them. To ensure that the franchisees perform as they and the Commission expect, they are trained effectively to operate and manage their business. Each new franchisee is expected to undergo a five-day small business management programme before they can operate any Commission equipment or business.

Activity brief

1 Evaluate the quality performance of the Lotteries Commission.
2 How would you appraise the developing quality culture?
3 Compare the approach of the Commission to managing data, compared to an organization you are familiar with.

Quality improvement tools and BOD Inc.: A football team

Background

The management of a leading football team — BOD Inc. — has had some major problems to contend with in the past three seasons. A succession of managers and coaches has led to an unfavourable working climate. Player contracts have been negotiated down from a football league average of 3.5 years, to an annual negotiated contract. In 1992, the club (the first) was launched onto the Stock Exchange amid great excitement and anticipation. However, declining positions, first in the Premier League, relegation last season, and now bottom of the First Division has resulted in support from the fans steadily declining, and the chairperson has sold all shares, and severed any further ties with the club.

The major concern for management is the on-the-field performances, as these determine what the club can achieve in the short term. Good performances will lead to increased support on the terraces, and this in turn will help satisfy both the bank manager and other stakeholders, such as shareholders. It is believed that if the club is relegated again, that their business position will become very serious.

Situation

The new caretaker manager has an interest in quality management, having attended a number of short seminars. The caretaker manager carried out a brainstorming session with all members of playing and support staff. The objective of the session was to determine:

> Why we can't win a football game?

This, therefore, has been taken as the ultimate objective of the exercise to ensure that:

1   All possible causes have been identified that may contribute to the team failures of the recent past.

2   These causes and their symptomatic effect have been analyzed as the resultant data would provide a means to indicate the general areas where improvements could best be made.

The brainstorming session results

A statement of general symptoms that supported the above issue was:

1   The team is not fast enough.
2   The team is not fit enough.
3   The team is not committed enough.
4   The Club management is not organized enough.
5   The team have not trained enough.
6   The team members have lower technical skills, education and abilities.
7   The team members have insufficient knowledge of the rules of the game.
8   Insufficient motivation of the team members.
9   The process of choosing the team members is suspect.
10 Insufficient capability of the support staff.
11 Coaching methods are not effective.
12 Insufficient knowledge of opposition tactics.
13 Insufficient facilities.
14 Insufficient time given for team development.
15 Amateurish attitude.
16 Level of player experience.

The next step

The manager does not now know how to proceed. The Secretary to the Club has suggested that a quality management consultant should be engaged in order to help focus the quality improvement effort.

Activity brief

1   As the quality management consultant, determine what charts and what additional data will be needed.
2   Develop a cause and effect diagram that answers — Characteristics of a Good Football Team.
3   Develop a cause and effect diagram on an identified problem area.
4   Describe what you anticipate are the problems of implementation of the resultant diagram from (2) above.

Problems

Problem No. 1

Acquire the 'organization for quality plan' of a service-oriented and a production-oriented organization. Compare and contrast their contents.

Problem No. 2

A manufacturer produces rubber inner tubes for bicycles. The critical dimension is the thickness of rubber at the air intake. The average for this has been calculated from over thousands of measurements as 0.75 mm, with a standard deviation of 0.15. Engineers have refined the manufacturing process and a test has resulted in an average of 0.79. Using the same standard deviation, if an increase in the dimension is required by designers, can the refined process be adopted?

Problem No. 3

Air Machinery Ltd

Air Machinery Ltd is attempting to secure a contract worth millions of pounds. In order to supply the components, and determine the lowest possible costs, the company is evaluating whether its present E4 production line can manufacture product within the specifications set by the prospective customer.

One critical feature is the chromium plating on the air damper, which must be within the specification regarding the thickness of 10 microns ±2 microns.

Air Machinery Ltd have responded by adapting the quality control system to monitor chromium thickness. A sample of four measurements has been taken from each of the last ten batches, which have been judged as typical production runs. The means and ranges of the ten samples were as follows (in microns):

Sample no.	1	2	3	4	5	6	7	8	9	10
Mean (x)	10.5	11.25	11.75	9.25	9.25	10.0	9.75	11.0	8.75	10.0
Range (R)	2.55	1.75	4.10	3.25	4.35	1.25	3.20	2.75	4.25	1.10

Construct X and R charts and comment on the picture that emerges.

Problem No. 4

Apply the TQM principles to the college that you attend. Discuss the problems and opportunities that may be encountered and achieved.

Problem No. 5

Compare and contrast the different approaches that may be used to implement TQM in a small and a large organization.

Problem No. 6

Outline Deming's fourteen points and apply them to an organization you are familiar with. What is the impact of their application?

Problem No. 7

Examine Juran's concept of the internal customer. Develop an analysis of how this might be implemented in an organization.

Problem No. 8

Examine the operation of a local small business. How would you manage the relationship between the various quality views?

Problem No. 9

Develop a quality mission statement and highlight the objectives you would develop for a new company that you anticipate setting up in the near future in a service industry.

Problem No. 10

One hundred samples of size 5 were taken from the production line of ball-bearings for the car parts industry. Each ball-bearing was measured and classified as acceptable or defective. The results indicate:

Sample no.	1	2	3	4	5	6	7	8	9	10
Defective %	5	2	7	3	1	5	8	2	1	4
Sample no.	11	12	13	14	15	16	17	18	19	20
Defective %	6	2	4	1	8	6	3	6	1	8

1. Plot a control chart showing the control limits.
2. Discuss any quality implications.

Problem No. 11

A new production line has just been commissioned and you have been asked to evaluate its quality-related performance. The designed specification for the measured process characteristic is 100 mm ± 12 mm. A preproduction run is used to formulate the data using ten samples, each of five units. The results of the run are:

Sample no.	1	2	3	4	5	6	7	8	9	10
	92	106	88	97	100	114	108	93	109	89
	95	111	103	101	103	106	103	95	101	85
	107	100	106	103	108	94	109	106	115	103
	104	95	98	96	99	92	105	109	114	107
	97	109	102	98	96	97	111	113	106	102

1 Develop X bar and R control charts for the above data.
2 What should be done next?

Problem No. 12

A manufacturer of plastic tubing has evaluated the process that produces it. The X bar and R control charts show that the process is in control, but defective product is still seen on final inspection.

1 What do you think can account for this?
2 What steps would you now take to rectify the problem.

Problem No. 13

A task force has been set up to investigate the quality problems on a production line, manufacturing plastic pen tops. Various problems have been raised by the quality department. These include:

1 The shipment of product that was non-conforming to the specifications set to an important customer. Engineering and production management overruled the quality department and sent the product to the customer.
2 A product lot was evaluated as having a small number of defective units. Since the product was to be used in a safety critical component, a debate ensued with production and marketing departments about whether to sort the lot for defectives — and then ship. The production manager wants to ship the lot straight away, as the manager indicates that it is a statistical problem, rather than a real problem.

You are the CEO and you need to resolve these problems immediately.

1 What would you do now?
2 What would you do in order to prevent these problems arising again?

Problem No. 14

A manufacturer of car body parts is concerned about the number of body panels found with paint imperfections after the hardening stage. This costs the company both time and costs in rework. Concerned about this, the CEO has asked the quality department to set up a monitoring programme and record the number imperfections. The result of the last twenty inspections reveals:

Sample no.	1	2	3	4	5	6	7	8	9	10
No. of imperfections	2	8	4	15	7	1	13	7	2	5
Sample no.	11	12	13	14	15	16	17	18	19	20
No. of imperfections	9	6	17	10	7	3	9	2	14	6

Compute and plot the appropriate control chart.

Problem No. 15

1. Develop a process flowchart for:
   (a) An examination in a department in an educational institution.
   (b) Producing a cup of coffee in a drinks vending machine.
2. Compare and contrast the differences and the likely and potential areas for error.

Problem No. 16

A routine internal quality audit in the marketing department has resulted in a number of major non-conformities that need to be addressed. Reviewing the audit schedules and non-conformance action statements, you find that the department consistently fails to adhere to the requirements of BS EN ISO 9000. This has put an intolerable strain on the registration commitment and could affect the whole certification process.

1. What do you think can account for this?
2. As the quality manager, advise as to what should be done next.
3. What steps would you now take to rectify the highlighted problems?

Problem No. 17

A process produces rubber oval balls. Measurements are taken on apex to apex, over fifteen samples size $n = 5$. The following shows the results:

Sample number	X̄	R
1	115	8
2	109	3
3	104	11
4	119	9
5	107	6
6	114	4
7	116	9
8	101	14
9	108	6
10	103	4
11	112	12
12	109	6
13	120	10
14	115	8
15	111	6

1 Set up X̄ and R charts on the rubber ball process.
2 If the process seems to be out of control, assume assignable causes and recompute the charts.
3 If the process is now in-control, reduce the computed specification by 10 per cent and recompute. What does the picture indicate?

Problem No. 18

A quality monitoring programme in a glass manufacturer's finishing section has resulted in the data shown as:

Sample number	Number of non-conformities
1	2
2	4
3	1
4	0
5	4
6	6
7	3
8	5
9	1
10	3
11	5
12	0
13	3
14	1
15	4

1 Set up a control chart for non-conformities.
2 Is the process in statistical control?
3 Comment on the picture that results.

Recent research

The results of a small business BS EN ISO 9000 survey in South Wales

Concerns raised in the piloting of the survey

Small businesses generally lack two major resources — human and capital. Consequently, any research study must have a direct impact on these very finite elements in order for them to be useful and effective. The BS EN ISO 9000 survey had to take this into account, not only through the process of inquiry, but also through the new data generated regarding the survey objectives.

The major aims of the study were to:

1   Determine the perceived status of importance of the BS EN ISO 9000 quality management standard in a sample of service-oriented small businesses.
2   Determine the status of the application of the various BS EN ISO 9000 elements.
3   Evaluate the possible quality management system-related impacts of a service focus.

A small business was determined as being a company that employed fewer than ten people.

Problems of the small business inquiry

One major problem, which needed to be attended to, was the sometimes academic nature of the nomenclature used. This was unavoidable, as the standard suggested the wording arrangements. Specifically, when surveying worker — rather than managerial — staff, some of the terminology used was deemed unfamiliar. Even though misunderstanding of questions may have led to irregular answer patterns, this could not be eliminated except by in-depth interviewing — and as such could assist the staff in developing a greater perception about the statements than may otherwise be the case when answering queries unassisted. This was minimized by conducting a seminar on the interpretation of the standard through which a set of explanations could be developed and used.

A second issue was caused by unfamiliarity with the Likert Scale. However, this was minimized by applying the use of a scale of 1–10, which assisted their greater recognition of the value of the units used. Thus, the use of this resulted in more reliable data. The population is very familiar with judging on a scale, of 1–10, by which 1 can be interpreted as *not important* and 10 as *very important*. On discussion with a pilot sample, this did not configure as a problem, as each ordinal point was precisely determined. This meant that it did not become necessary, even during the pilot, to change the scale used. The results of the actual surveys did not lose their validity in respect of the scale adopted, as the results were only compared with each other.

A third issue was the time required for staff to answer the survey statements. The time staff made available was often very short. This creates two separate problems: firstly the dependability of the results, as to their perception of quality, and secondly the amount of statements for staff to consider needed a commitment of time, not always appreciated by the staff surveyed. Experience with the pilot clearly indicated that relevant information could be extracted by using just a survey.

A fourth issue was that staff did not seem to accept that quality of service was a necessity in all services provided. This may be a major misgiving of small businesses, whose cultural orientation was more processor, rather than customer.

Methodology

A sample of 150 small companies was randomly selected with a clear service orientation. The geographical spread covered the major towns and cities found in South Wales. The population frame also covered both inner city and rural areas.

Procedure

The managing director or company owner was contacted by telephone, in order to ascertain the likelihood for co-operation in the quality study. However, even at this juncture, it became clear that many directors ($n = 35$) were not altogether interested in participating further, as it would mean taking too much time to do the evaluation, or that the results would indicate the inadequacy of their working practices. This in itself is significant, as some small companies may not want to know the effectiveness of the processes or staff that produce their products or services. This left 115 companies for evaluation.

Response rate of survey

A written survey was sent to each director or owner indicating due date, and this was followed by a telephone call to enquire about potential problems. By the due date, 79.13 per cent (91) had returned completed questionnaires. This high response is taken to reflect some of the pressures on small companies due to the quality revolution.

Findings

Table 1 indicates the result.

Table 1 ISO 9000 findings — service-oriented companies ($n = 91$)

	Statement	Importance rating (mean)	Rank
1	Ensuring that the policy for quality is understood, implemented and maintained at all levels in the organization is	8.20	8
2	Defining the responsibility, authority and interrelation of all personnel who manage, perform and verify work affecting quality is	9.20	1
3	The identification of and resource of verification requirements is	1.40	16
4	The appointment of an individual to manage the requirements of BS EN ISO 9000 so that they are implemented and maintained is	7.80	9
5	Management ensures the continuing suitability and effectiveness of the quality system is	9.10	3
6	Documents that constitute the quality system are employed to ensure that the product conforms to specified requirements is	6.90	12
7	Controlling and verifying product/service design is	1.90	15
8	Controlling documents and data that relate to the requirements of BS EN ISO 9000 is	8.70	5
9	Identifying product in a way that indicates the conformance or non-conformance with regard to inquiries, inspections or examinations performed is	7.40	11
10	Ensuring that product that does not conform to specified requirements is prevented from inadvertent use is	6.40	14
11	The detection and elimination of potential causes of non-conformance of product and prevent their re-occurrence is	7.60	10
12	Prevention of damage or deterioration to product in handling, storage and delivery is	7.90	8
13	The verification of quality activities that comply with planned arrangements and the determination of the effectiveness of the quality system is	9.20	1

Table 1 continued

Statement	Importance rating (mean)	Rank
14 The identification of training needs that ensure that personnel performing specific assigned tasks are qualified on the basis of appropriate education, training and/or experience is		6
15 Ensuring that service delivery is performed and verified in a way that meets the specified requirements is		4
16 The identification of adequate statistical techniques required for verifying the acceptability of process capability and product/service characteristics is		13

The BS EN ISO 9000 statements were thought to be of some importance when assessing quality. Only one statement was seen as irrelevant by less than 3 per cent of the respondents. Fourteen were found to be essential by more than half (54 per cent) of the respondents, and of these 66 per cent of the statements were answered at 7.0 or above.

As expected, there was wide variation seen in the extent to which respondents regarded statements as essential, ranging from 78 per cent for 'ensuring that service delivery is performed and verified in a way that meets the specified requirements', down to just 14 per cent of the respondents who rated 'the identification of adequate statistical techniques required for verifying the acceptability of process capability and product/service characteristics'.

In one respect, it is not surprising to find that the majority of statements were rated as they were. Some respondents had prior opinions about BS EN ISO 9000, but many did not. Thus, the response may be a reflection of hearsay, rather than through informed attempts deliberately to provide an effective opinion. This has implications for managing quality in small businesss and especially for the application of BS EN ISO 9000, beyond even BS EN ISO 9003.

Some statements were deemed rather poorly evaluated, e.g. statement 7, which reflected procedures for controlling and verifying product/service design scored a reasonably low score of 1.9. This suggested that the actual application of the statement was not understood effectively, or that the statement was not actually required in their view. However, this considered, small businesses who do not carry out this in the normal course of their business risk not satisfying their customers, and therefore losing them to the competition.

Statement 3 was also poorly scored (1.4) and perhaps indicates that low scores could represent, not the lack of importance, *per se*, but lack of understanding of the importance of its application, since much of the standard is in the language of industry, rather than services.

Statement of outcomes of the ISO 9000 research

The perceived importance of many factors and their application in reality are very far apart. Particular problems are deemed to exist in the application of management areas, e.g. objective setting, planning and leadership.

Staff issues are also highlighted as the importance rating (8.3) determines. This contrasts with statement 14 (score 8.35) that immediately gives greater importance to training requirements. However, the evidence suggests that this is rather poorly supported in reality. This may be because of inadequate funding, government support or that margins do not provide the profit necessary to invest in staff appropriately. Quality review requirements also hold a high importance of up to 90 per cent. Again, procedures in the department seem to be perceived as inadequate and do not provide for effective service delivery, nor do they provide for confidence in and of, the staff.

Conclusions

Although great significance has been placed on the perceived importance by small businesses in South Wales, this BS EN ISO 9000 survey has only indicated that much help must be given to small businesses if they are to succeed against more resourced, much larger market opposition. The survey also indicates that BS EN ISO 9000 interpretation is biased towards larger organizations, thus creating an unfair barrier to entry to the market place. This is especially so when it is seen that nearly 40 per cent of customers of small businesses are large firms. However, more than 60 per cent of small businesses are customers of other small businesses, and it is here that much of the wealth is determined. Consequently, much work must be done to ensure that small businesses are given the support[1] and resources necessary to ensure their attachment to the quality revolution — otherwise customers, clients, small and large businesses will fail each other. This can only mean a serious detriment to the basis for a quality orientation — that of enhanced competition and ultimately survival.

1. The IQA (UK) are carrying out a process of evaluation of BS EN ISO 9000 and the small business.

Quality glossary

Ability One measure of the application of an individual's skill, competence or proficiency.
Acceptance sampling Auditing a sample from a given lot size, and determining from the result of counted defectives whether or not to accept the whole lot.
Actions Specific, definable means to accomplish goals or objectives.
Affiliation needs The generated needs that reflect friendship and social requirements.
Analytical skills The ability to use methods or techniques in order to solve problems.
Auditing The process of determining whether a quality management system is peforming to expected standards.
Authority The right of an individual to make decisions and expect others to carry out given tasks.
Avoidance style The action designed to withdraw from a perceived situation of conflict.
Behavioural model A leadership model that focuses on behaviour to determine the best means of leading in a given situation.
Benchmarking The process of evaluating an organizational process, or practices against another process or practice within or outside of an organization.
Brainstorming The process where individuals furnish ideas, notions or comments about a given problem or opportunity, without having ideas, notions or comments prejudged.
Breakthrough A defined improvement that departs from previous practices, accomplished through creativity and innovation.
Bureaucracy Methods of administrative control that formulate structured or rigid organizations.
c chart A control chart that represents either the total number of non-conformities in a unit or the average number of non-conformities found per unit.
Cause and effect diagram A graphical representation of the present ideas, notions or thoughts on the cause of a given problem.
Centralization The process of retaining power and authority in top management and those who make all strategic decisions.
Chain of command The lines of formal authority that extend hierarchically throughout the organization from top to bottom.
Change agent An individual or individual representing a group, who is charged with suggesting or enforcing change on individuals or groups in an organization.
Closed system A system that has no interface with another external system.
Coaching style A leadership style that is characterized by a manager/leader using two-way communication to help subordinates develop more effectively.
Coercive power Power given to a leader by a subordinate by virtue of fear of punishment through non-compliance.
Cohesiveness A group process that results in greater individual commitment to the objectives of the group.
Collaborative style A style of conflict that results in the development of shared problem-solving approaches where both or all parties gain.
Common cause of variation The random, erroneous variation contained in a system.
Communication The process of transferring meaning between two or more individuals.
Communication feedback The intended receiver's response to the sender's message.

Communication skills The ability to create understanding through the communication process.
Compensation The remuneration in the form of wages, salary, stocks or shares that constitute a return of value to an individual for doing work.
Competitiveness The degree to which goods and services successfully and continuously meet customers' needs and wants.
Compromise style A style of conflict where two or more individuals sacrifice part or all of their position or interest in order to develop a solution to a problem.
Concept to customer The time taken to design, produce and sell goods and services to the final consumer.
Conceptual skills Abilities that reflect an holistic approach to planning and organizing activities.
Concurrent control Simultaneous monitoring of production process in order to ensure that objectives are pursued.
Conflict This is caused through differences in individual perspective arising in groups or organizations.
Conflict management Structural interventions designed to manage identified conflict situations effectively.
Conformity The degree to which individuals in a group match the development group norms, roles and behaviours.
Content theories A category of motivation theory that focuses on the attainment of inner needs.
Contingency models Leadership theories that suggests that the external environment must be considered when choosing a leadership style.
Contingency planning Planning that involves developing strategies that relate to every identified alternative.
Continuous improvement The positive direction taken by applying incremental steps to streams of adaptive decisions, over a period of time.
Control Mechanisms designed to ensure that set objectives are met according to specifications.
Control chart A graphical record of measured process results, over a period of time, showing designed or developed upper and lower control limits.
Controlling The process by which an individual exerts conscious decisions to effect any necessary corrective actions. One of the management functions.
Co-ordination The formal (and informal) processes used to assure that interdependent groups function together.
Cross-functional system A form of structure involving a number of functions and used to provide an integrated and co-ordinative approach to ensure products and services satisfy customers.
Culture Shared characteristics of behaviour, dress and language that distinguish one group or organization from another.
Customer Final consumer or recipient of the result of the operation of a production process.
Data Facts and figures that have no inherent meaning.
Data processing Actions or processes designed to transfer data into information.
Decentralization This is characterized by delegating authority to subordinates; the more delegated the authority, the more decentralized the department or organization.
Decision making The process of recognizing a problem, developing and evaluating alternatives and choosing one.
Decision-making process A related series of organized steps leading to a decision.
Decisional roles Roles characterized by the management of an organization as decision makers.
Decoding The process of translating a message into meaningful information.

Delegating style   A leadership style characterized by the deployment of responsibilities to subordinates.
Depth of intervention   The degree of change an intervention is designed to bring about.
Descriptive statistics   A computed measure of a data set that provides the basis for statements regarding its specific meaning.
Design strategy   The organizing methods of ensuring that customer needs (internal and external) are met now and in the future.
Directive behaviour   The action of one-way communication between leader and subordinate.
Directive style   A leadership style that uses the medium of one-way communication to execute instructions or commands.
Disseminator role   The informational role that managers use to share information inside the organization.
Division of labour   The process of breaking down jobs into smaller tasks in order to increase efficiency.
Downward communication   Information flows that move downward through an organization.
Employee involvement   A process by which subordinates are encouraged to make decisions at their level in the hierarchy.
Empowerment   Authority given to subordinates, who are able to make certain decisions without asking for prior approval.
Encode   The process of translating a message into a form that is perceived to be meaningful information to the receiver.
Environmental analysis   The action and methods of scanning both internally and externally in order to determine strengths, weaknesses, opportunities and threats.
Ethics   Personal behaviour that is characterized as a result of an individual's belief and value system.
Expectancy   This relates to the degree of probability that an individual's efforts will lead to satisfactory task performance or reward.
Extinction   The simple process of non-reinforcement, over time, of a given behaviour by ignoring its substance and results.
Extrinsic rewards   Rewards that are seen as externally controlled.
Feedback   The receiver's response to the sender's interpreted message.
Feedback control   Specific control measures designed to alter the input conditions of a process positively through end-line monitoring.
Fiedler's LPC theory   A two-step theory of determining a leader's style, then finding a given situation that is conducive to a test and applying that leader's style.
First-line management   Managers normally known as supervisors, characterized as technically oriented, rather than administratively oriented.
Forecast   A structured prediction of future events or occurrences.
Formal group   A group that is characterized whose existence, roles and tasks are sanctioned by the organization.
Formal training programme   A formal effort made by an organization to provide job-related training, skills and knowledge.
Formalization   The degree to which an organization's procedures are written and made available to staff.
Forming   The first stage in the development process of a group.
Goal   A defined performance target developed in advance.
Goal-setting theory   This is characterized by goal-setting individuals outperforming other individuals who do not set goals.
Grapevine   Informal organizational communications that transcends the formal communication channel with opinions, facts and rumours.

Group Two or more who become identified as engaged in collective activities directed towards the same goal.
Group norms Collectively developed sense of behaviour standards.
Groupthink Internally developed actions that do not test outside reality.
Hawthorne effect The tendency of changes in normal behaviour of the observed.
Hawthorne studies Research studies conducted by Mao in Western Electric's Hawthorne plant in Chicago, USA.
Hierarchy An organizational structure characterized by the division of work into specialized departments and functions.
Hierarchy of needs A middle-class motivation theory developed by Maslow — physiological, safety, social, self-esteem and self-actualization.
Histogram A graphical illustration of the distribution pattern of data or measurements.
Horizontal communication Information that flows between individuals on the same level in an organization.
Human resource management The process of selecting, recruiting, training and rewarding individuals in the accomplishment of organizational goals.
Human resource planning The process of determining the organizational human requirements by estimating the size and needs of the future workforce.
Hygiene factor The aspects of Hertzberg's two-factor theory that relate to the external environment, rather than to work itself.
Implementation The application of effort to achieve a set objective or plan.
Informal group Two or more individuals involved in voluntary activities that are directed to a common goal solution.
Information Data that have been organized and analyzed to provide meaning.
Information processing The sequence of steps designed to convert data into information. That is, meaningful for decision makers.
Informational roles Roles assumed by managers to receive, send and control information flow in an organization.
Innovation The creation and development of new ideas, products or processes that provide a basis for competitive advantage.
Inseparability An environment where services are produced and consumed simultaneously.
Inspection An examination whose purpose is to attempt to measure certain characteristics or to identify defects.
Intangibility A characteristic or attribute that resists assessment.
Internal customers Discrete internal processes, whose output is subject to the next determined customer requirement, and thus form a chain of satisfaction.
Interpersonal communication Information exchange between two or more people.
Interpersonal roles The roles assumed by managers that enable them to perform informational and decisional roles.
Intervention A method or technique used to assess the need for, or change the structure, behaviour or the technology use of a group or organization.
Job analysis The process of evaluating the content and requirements of jobs and their related tasks.
Job depth The degree to which an individual has discretion over task performance.
Job description A written summary of a job's characteristics, that details responsibilities, tasks and operational conditions.
Job design The articulated design for the performance of the job content and tasks.
Job enlargement Increasing the number of tasks allotted to an individual job holder.
Job enrichment Increasing the degree of control an individual exercises over a specific job content that addresses the need for individual growth and development.
Job rotation Planned movement of an individual from one job to another.

Job specialization The process of breaking a job down into smaller discrete tasks for greater efficiency.
Job specification A written description of the skills, competencies and knowledge required of the job holder.
Juran's trilogy An integrating approach that is characterized by the use of three managerial processes of planning, control and improvement.
Just-in-time (JIT) A process of ensuring that raw material/supplies are made available at the time they are required.
Leadership The process of exerting influence over an individual.
Leading One of the management functions that seeks to ensure, through positive influence, that subordinates carry out organizational tasks effectively.
Learning The process by which individuals change their attitudes and behaviour to acquire skills, knowledge and abilities.
Linking pins These are characterized by a manager being a leader in one hierarchical group, and a subordinate in another.
Management The process of planning, organizing, leading, controlling and staffing that lead to effective task solution.
Management level The right to direct individuals according to the application of the vertically specialized management process.
Managers Individuals who direct, control and sanction other individuals in the administrative performance of managerial tasks.
Manufacturing The process of producing tangible product.
Manufacturing-based view of quality A view that is strategically focused on operations through the emphasis and orientation to conformance to design specification.
Mass production The process of producing large quantities of product in a given timeframe, generally resulting in economies of scale.
Matrix organization A cross-functional structure that results in greater horizontal co-ordination of specialized tasks and responsibilities.
Mechanistic organization A traditional organization that produces efficiency through the development of features of centralized authority, rules and highly specialized jobs.
Medium of transmission A means of carrying an encoded message from the sender to the receiver.
Middle management Managers who manage the input/output of other managers, and who themselves are subject to managerial influence.
Mission statement A statement that creates the vision, sense of purpose and direction for an organization.
Motivation Forces (internal and external) that determine reactive behaviour to that stimuli.
Negative reinforcement Reinforcement that occurs when an unpleasant consequence is withdrawn each time a desired behaviour occurs.
Noise Interference that upsets the meaning given to the intended message between sender and receiver.
Non-conformance Exhibited by the identified non-fulfilment of a specified requirement.
Norming The stage in the group lifecycle when the group develops a code of conduct for its operation.
Objective A target that must be strived for, and accomplished.
Open system A system that interacts with its external environment and uses feedback in order to make changes or modifications to its operation.
Operations The central functions of performance in an organization.
Organic organization This is characterized by being flexible and flat.
Organization A set of interdependent and interrelated parts that require co-ordination mechanisms in order to control performance and attain set objectives.

Organizational design The process of developing organizational structure.
Organizational structure The developed framework of jobs, departments and functional disciplines that direct individual behaviour towards organizational goal accomplishment.
Organizing The management function that assigns tasks and responsibilities to individuals.
p **chart** A control chart of a type that graphically illustrates variation in the fraction nonconforming of a production process.
Pareto analysis A graphical representation of a bar graph and accumulation line of data events organized in reducing order.
Path–goal leadership theory Forms part of the expectancy theory of motivation in which an individual engages in organizational goal accomplishment that leads to the generation of rewards that the individual values.
Performance evaluation A review of individual performance by using post-control techniques.
Performance standards Expected outcomes as previously defined and set.
Performing The stage in the group lifecycle when the group functions at its best potential.
Perishability Characterized as indicating that service capacity cannot be stored for later use.
PDCA cycle The four-step process of plan, do, check, act cycle of Deming.
Planning The primary function of management that determines what needs to be achieved, by when and with what.
Policy A written statement that broadly determines guidelines for action to achieve set goals.
Positive reinforcement The process that provides rewards for the production of expected behaviour.
Power The operation of overt and covert influences in an organization.
Problem An identified state that indicates a discrepancy between what is expected and the current state.
Problem solving Attempts at locating the real cause, and developing solutions to that cause, so that the cause does not return.
Procedure A written clause, paragraph or document that describes what is to be carried out, why, by whom, when and how.
Process The flow of materials, information and product that have been organized in advance, in order to translate inputs to expected output.
Process improvement The deliberate act of monitoring a process outcome and developing ways to enhance its future performance.
Product-based view of quality The view that quality is encompassed in the totality of features and characteristics of the manufactured product.
Production The overall process that produces goods and services.
Punishment The administration of an undesirable consequence in response to undesirable behaviour.
Quality The notions or characteristics that formulate an opinion in regard to a product or service and can be expressed in terms of the psychological, product or manufacturing process; the totality of features and characteristics of a product or service that bear on its ability to satisfy implied or stated needs (ISO 8402).
Quality assurance The development of a quality orientation, that uses internal processes to assure the product meets stated specifications by focusing on prevention.
Quality audit A systematic and independent evaluation of a process to determine whether the process meets the planned set standards in terms of consistency and output characteristics.

Quality circle A voluntary group of individuals (up to ten), who come together in order to identify problems related to their area of work.

Quality function deployment (QFD) A methodology that transfers the customer's voice through the organization by using a chain of graphical matrices.

Quality manual A written or on-line computer document that describes the basis for the quality mission, policy and developed procedures and quality practices that an organization has determined should be used in a quality management system.

Quality planning The process of determining and developing plans that create the direction for the organization.

Quality system The organizational structure, responsibilities, procedures and resources for implementing quality management (ISO 8402).

Quality system review A regular, structured and formal examination of the quality system in order to determine the adequacy of the implementation of the quality policy in order to make positive quality-related improvements.

R chart Control charts that represent the spread of measured variation of a characteristic over a period of time, and that also indicate the statistically calculated boundaries or limits of control.

Recruitment The process of attracting job candidates with the prescribed abilities, knowledge and skills to assist the organization attain its objectives.

Reinforcement The process of encouraging future desired behaviour.

Role Behaviour patterns, responsibilities and performance requirements expected from an individual holding a given post or position.

Scientific management Efficiency oriented practices, developed by Taylor.

Scope The range of activities covered by a given plan.

Selection The process of choosing an individual or individuals who best meet set criteria for an available position in an organization.

Self-leadership The practice of allowing individuals to reach for higher performances by expressing greater personal influence over own job activities.

Self-management The process of managing own work requirements without external assistance.

Sender The encoder and originator of a message.

Service A predominantly intangible product.

Service quality Characterized by the degree to which the service conforms to express or implied customer requirements.

Seven new tools of quality Advanced engineering methods for developing processes more effectively.

Seven old tools of quality Structurally simple methods used to analyze data generated in a process.

Single-use plans Plans that are predominantly project based, with a clearly defined timeframe and operation.

Situational theory Leadership theory that reflects the need to change as the environmental context changes.

Skill A proficiency in performing a particular practice.

Span of control The number of individuals who regularly report to a manager.

Special cause of variation Variation in a system that can be identified and accounted for.

Stable system A system that is considered to be in statistical control through the elimination of special cause variation.

Standing plans Routine plans that have become accepted as part of the standing operating procedures of the organization.

Statistical control limits Limits calculated through the use of statistical methods, applied to a series of data, that represent the boundaries for a stable variation system.

Statistical process control Methods designed to ensure that a process attains and remains in a state of statistical control.

Storming The stage of development of a group lifecycle when attention is given to solving internal conflict.
System Interdependent parts that operate collectively to achieve common goals.
Team A group of individuals who have a common goal, and are identified as related in some way.
Team building An organizational development intervention that seeks to improve the communication, co-operation and performances of individuals in groups.
Technical skills The ability to use specific job-related knowledge, techniques and other resources to accomplish work tasks.
Technological change The planned and systematic change of the technologies used by an organization.
Technology The means used to develop, process and deliver products and services in order to satisfy customers.
Total quality management An ideologically culturally based system of managerial operation that seeks to improve continuously the total organizational system that produces goods/services to satisfy customers every time.
Training The systematic process of changing an individual's job-related behaviour.
Trait theory A leadership theory that focuses on the leader's physical and psychological attributes.
Transcendent view of quality View of quality that is highly personal and resists definition, as it is deemed to be an unanalyzable property resulting from personal taste or style.
***u* chart** A control chart used to represent graphically variation in whole counts of events.
Upward communication Information that flows upward through the organization from subordinate to manager.
User-based view of quality View of quality that results from the use of the product or service.
Variable data Data produced from measuring a characteristic feature of a product or service at regular intervals.
Variation Changes in data value received from a process over time.
Vision A deliberate and clear understanding of an organization's future.
X bar charts Control charts that represent the measured variation of a characteristic over a period of time, that also indicate the statistically calculated boundaries or limits of control.
Zero defects A concept where no product/service defects are acceptable.

Bibliography

ABED and DALE 'An attempt to identify quality-related costs in textile manufacturing', *Quality Assurance*, 13 (2), 1987
ADAIR *Training for Communication*, Macdonald, 1973
ADAIR *Effective Leadership: A Self-development Manual*, Gower, 1983
ALBRECHT and ZEMKE *Service America*, Irwin, 1985
ALDERFER *Existence, Relatedness, and Growth: Human Needs on Organisational Settings*, Free Press, 1972
ANON 'An American tragedy', *The Economist*, August 22, 56–58, 1992
ATKINSON *Creating Culture Change: The Key to Successful Total Quality Management*, IFS Publications, 1990
BARKDULL 'Span of control: a method of evaluation', *Michigan Business Review*, 15, 1963
BARROW 'The variables of leadership: a review and conceptual framework', *Academy of Management Review*, April, 1977
BARTOL and MARTIN *Management*, McGraw-Hill, 1991
BASS *Leadership, Psychology and Organisational Behaviour*, Harper and Row, 1965
BASS *Stogdill's Handbook of Leadership: A Survey of Theory and Research*, Free Press, 1981
BEER *Organisational Change and Development*, Goodyear, 1980
BEMOWSKI 'The competitive benchmarking wagon', *Quality Progress*, Jan., 1991
BERGER and LUCKMAN *The Social Construction of Reality*, Penguin, 1967
BERNE *The Structure and Dynamics of Organisations and Groups*, Grove Press, 1963
BION *Experiences in Groups*, Basic Books, 1959
BLAKE and MOUTON *The Managerial Grid*, Gulf Publishing, 1964
BOGAN and ENGLISH 'Benchmarking: a wakeup call for board members', *Planning Review*, July/Aug., 1993
BOWEN and SCHNEIDER 'Boundary-spanning-role employees and the service encounter', in Czepiel, Solomon and Suprenant (eds), *The Service Encounter*, Lexington Books, 1988
BRELIN 'Benchmarking: the change agent', *Marketing Management*, 2 (3), 1993
BROCKA and BROCKA *Quality Management: Implementing the Best Ideas of the Masters*, Business One Irwin, 1992
BROWN *The Social Psychology of Industry*, Penguin, 1967
BUCHANAN and HUCZYNSKI *Organisational Behaviour: An Introductory Text*, Prentice Hall, 1985
BURKE *Connections*, Macmillan, 1978
BURNS *Leadership*, Harper and Row, 1978
BURNS and STALKER *The Management of Innovation*, Tavistock, 1961
BUSH *Managing Education: Theory and Practice*, Open University Press, 1992
CALDWELL 'Cultivating human potential at Ford', *The Journal of Business Strategy*, Spring, 1984
CAMP *Benchmarking: The Search for Industry's Best Practices That Lead to Superior Performance*, ASQC Quality Press, 1989
CARROL and GILLEN 'Are the classical management functions useful in describing managerial work?', *Academy of Management Review*, 12, 38–51, 1987

CEC Draft Publication *Total Quality Management in Education and Training*, Brussels, January, 1992
CHAPMAN 'Investing in people for business success', *Quality World*, Oct., 1994
CHASE 'Where does the customer fit in a service operation', *Harvard Business Review*, 56, Nov.–Dec., 1978
CHASE and TANSIK 'The customer contact model for organisational design', *Management Science*, 29, 1983
CHILD *Organisation: A Guide to Problems and Practice*, Harper and Row, 1977
CHIN and BENNE 'General strategies for effecting changes in human systems', in Bennis, Benne and Chin (eds), *The Planning of Change*, Holt, Rinehart and Winston, 1979
CHURCHMAN *The Systems Approach*, Delta, 1968
Consumer Network 'Brand quality perceptions', *The Consumer Network*, Philidelphia, PA, Aug., 1983
COOK 'TQM research', *Proceedings of the 4th International Conference. Total Quality Management*, IFS Publications, June, 1991
COUBROUGH *Fayol, Industrial and General Administration*, Trans. Geneva International Institute, 1930
CROSBY *Quality is Free*, Mentor, 1979
CROSS *Engineering Design Methods*, Wiley, 1989
CROW and VAN EPPS 'Competitive benchmarking: a new medium for enhancing training and development effectiveness', *Training and Management Development Methods*, 7 (1), 1993
CURTIS 'The quality partnership: responsibility and involvement', *Proceedings of the 2nd International Conference on Total Quality Management*, IFS Publications, June, 1989
DAKIN 'How to identify and assess human potential: The place of psychological testing', *New Approaches to HR Development Conference*, Wellington, New Zealand, May, 1989
DALE and PLUNKETT *Quality Costing*, Chapman Hall, 1991
DEMING *Quality, Productivity and Competitive Position*, MIT, 1982
DEMING *Out of the Crisis*, MIT, 1986
DEMING 'Leading quality transformations', *Executive Excellence*, 10 (5), May, 1993
DENBURG and KLEINER 'How to provide excellent company customer service', *International Journal of Contemporary Hospitality Management*, 5 (4), 1993
DESSLER *Management Fundamentals*, Prentice Hall, 1985
DRUCKER *The Practice of Management*, Harper and Row, 1954
DRURY *Management and Cost Accounting*, Chapman, 1992
DUNCALFE and DALE 'How British industry is making decisions on product quality', *Long Range Planning*, 18 (5), 1985
DURGESH and EVANS 'Effective benchmarking: taking the effective approach', *Industrial Engineering*, 25, February, 1993
EUREKA and RYAN *The Customer-Driven Company*, American Supplier Institute, 1988
EVAN *Conflict and Performance in R & D Organisations*, 1965
FEIGENBAUM *Total Quality Control*, McGraw-Hill, 1991
FEIGENBAUM 'Feigenbaum's window on the world: regaining the quality service edge', *National Productivity Review*, 12 (4), 1993
FIEDLER *A Theory of Leadership Effectiveness*, McGraw-Hill, 1967
FIEDLER 'The contribution of cognitive resources to leadership performance', *Journal of Applied Social Psychology*, 16, 1986
FOLLETT *Dynamic Administration: The Collected Papers of Mary Parker Follett*, Metcalf and Urwick (eds), Pitman, 1941
FORD 'Benchmarking HRD', *Training and Development*, 47 (6), June, 1993
FORD and RANDOLPH 'Cross-functional structures: a review and integration of matrix

organisation and project management', *Journal of Management*, June, 1992
FRASER-ROBINSON *Total Quality Marketing*, Kogan Page, 1991
FRENCH and BELL *Organisational Development*, Prentice Hall, 1984
FRENCH and RAVEN 'The bases of social power', in Cartwright (ed), *Studies in Social Power*, Institute of Social Research, Michigan, 1959
FRIEDMAN *Industry and Labour*, Macmillan, 1977
FULLAN *The New Meaning of Educational Change*, Cassell Education, 1992
GARVIN 'Quality on the line', *Harvard Business Review*, Sept.–Oct., 1983
GARVIN 'Competing on the eight dimensions of quality', *Harvard Business Review*, Nov.–Dec., 1987
GARVIN *Managing Quality*, The Free Press, 1988
GRABER, BREISCH and BREISCH 'Performance appraisal and Deming: a misunderstanding', *Quality Progress*, June, 1992
GREINER 'Patterns of organisational change', *Harvard Business Review*, May–June, 1967
GREINER 'Evolution and revolution as organisations grow', *Harvard Business Review*, July–Aug., 1972
GRIFFIN *Management*, Houghton Mifflin, 1993
GRYNA 'Quality costs: user vs manufacturer', *Quality Progress*, June, 1977
GUEST 'Of time and the foreman', *Personnel*, May, 1956
HACKMAN 'The design of work teams', in Lorsch (ed), *Handbook of Organisational Behaviour*, Prentice Hall, 1987
HACKMAN and OLDHAM *Work Redesign*, Addison-Wesley, 1980
HAMPTON 'Gap analysis of college student satisfaction as a measure of professional service quality', *Journal of Professional Services Marketing*, 9 (1), 1993
HASEROT 'Benchmarking: learning from the best', *CPA Journal*, October, 1993
HARVEY and BROWN *An Experiential Approach to Organisation Development*, Prentice Hall, 1988
HAUSER and CLAUSING 'The house of quality', *Harvard Business Review*, May–June, 1988
HEQUET 'The limits of benchmarking', *Training*, 30 (2), February, 1993
HERSEY and BLANCHARD 'Life-cycle theory of leadership', *Training and Development Journal*, 20 (2), 1969
HERSEY and BLANCHARD 'So you want to know your leadership style?', *Training and Development Journal*, February, 1977
HERSEY and BLANCHARD *Management of Organisational Behaviour: Utilising Human Resources*, Prentice Hall, 1978
HERTZBERG *Work and the Nature of Man*, World Publishing, 1966
HERTZBERG 'The manager's job: folklore and fact', *Havard Business Review*, July–Aug., 1975
HICKMAN and SILVA *The Future 500: Creating Tomorrow's Organisations Today*, Unwin, 1987
HOFSTEDE *Culture and Organisations: Software of the Mind*, McGraw-Hill, 1984
HOUNSHELL *From the American System to Mass Production (1800–1832)*, Johns Hopkins Press, 1984
HOUSE 'A path–goal theory of leader effectiveness', *Administrative Science Quarterly*, September, 1971
HOUSE and DESSLER 'A path–goal theory of leader effectiveness: some post hoc and a priori tests', in Hunt and Larson (eds), *Contingency Approaches to Leadership*, Southern Illinois Press, 1974
HOWELL and FROST 'A laboratory study of charismatic leadership', *Organisational Behaviour and Human Decision Processes*, 43, 1989
IRVINE *Improving Industrial Communication*, Gower, 1970
ISHIKAWA *What is Total Quality Control? The Japanese Way*, Prentice Hall, 1985
IVANCEVICH, LORENZI, SKINNER and CROSBY *Management: Quality and Competitiveness*,

Irwin, 1994
JAMES *Testing the Use of the Five Functions of Management at Dynamics Ltd*, unpublished research, 1994
JANIS *Victims of Groupthink: A Study of Foreign Policy Decisions and Fiascoes*, Houghton Mifflin, 1972
JOHNSON and SCHOLES *Exploring Corporate Strategy*, Prentice Hall, 1993
JURAN *Managerial Breakthrough*, McGraw-Hill, 1964
JURAN *Quality Control Handbook*, McGraw-Hill, 1974
JURAN 'The quality trilogy: A universal approach to managing quality', *Quality Progress*, August, 1986
JURAN *Planning for Quality*, Free Press, 1988
JURAN *Management of Quality*, Juran Institute, 1989
JURAN and GRYNA *Quality Planning and Analysis*, McGraw-Hill, 1993
KANTER *The Change Masters*, Unwin Hyman, 1983
KAST and ROSENZWEIG 'General systems theory: applications for organisation and management', *Academy of Management Journal*, 15 (4), Dec., 1972
KAST and ROSENZWEIG *Contingency Views of Organisation and Management*, Science Research Associates, 1973
KAST and ROSENZWEIG *Organisation and Management: A Systems Approach*, McGraw-Hill, 1974
KAST and ROSENZWEIG *Organisation and Management: A Systems Approach*, McGraw-Hill, 1979
KATZ 'Skills of an effective administrator', *Harvard Business Review*, Sept.–Oct., 1974
KEEN 'Adding value through people', *Training Tomorrow*, June, 1993
KOONTZ 'The management theory jungle revisited', *Academy of Management*, 10 (1), July, 1978
KOONTZ, O'DONNELL and WEIHRICH *Essentials of Management*, McGraw-Hill, 1986
KOTLER *Marketing Management: Analysis, Planning, Implementation and Control*, Prentice Hall, 1991
KOTTER *The General Manager*, Free Press, 1982
LABOVITZ and CHANG 'Learn from the best', *Quality Progress*, May, 1990
LETTERS
'Feel the width of total quality', *Times Higher Education Supplement*, July 19, 1991
'Ministers warned off 'Kite-mark' for colleges', *Times Higher Education Supplement*, July 12, 1991, p.1
Times Higher Education Supplement, August 2, 1991
'Confusion in search of total quality', *Times Higher Education Supplement*, Nov. 29, 1991
'Standards' respect due to respect', *Times Higher Education Supplement*, August, 1991
LEVACIC *Financial Management in Education*, Open University Press, 1993
LEWIN 'Frontiers in group dynamics: concept, method and reality in social science', *Human Relations*, 1, June, 1947
LEWIN *Field Theory in Social Science*, Harper and Bros, 1951
LIKERT *New Patterns of Management*, McGraw-Hill, 1961
LOCKE, SHAW, SAARI and LATHAM 'Goal setting and task performance: 1969–1980', *Psychological Bulletin*, 90, 1981
LONG 'Changing role of corporate planning', *Executive Excellence*, 10 (6), June, 1993
LOVELOCK 'Why marketing management needs to be different for service', in Donnelly and George (eds), *Marketing of Services*, AMA, 1981
LUTHANS 'The contingency theory of management: a path out of the jungle', *Business Horizons*, 16 (3), June, 1973
MAIN 'How to steal the best ideas around', *Fortune*, October 19, 1992
MAKIN and ROBERTSON 'Management selection in Britain: a survey and critique', *Journal of Occupational Psychology*, 59, 1986

MANZ and SIMS *Superleadership*, Berkeley, 1990
MARSHALL 'EuroView', *Quality World*, January, 1994
MASLOW 'A Theory of human motivation', *Psychological Review*, 50, 370–396, 1943
MASLOW *Motivation and Personality*, Harper and Row, 1954
MCCARTHY *Basic Marketing: A Managerial Approach*, Homewood, 1961
MCCLELLAND *Human Motivation*, Scott, Foresman, Glenview, 1985
MCGREGOR *The Human Side of Enterprise*, McGraw-Hill, 1960
MEFFORD 'Improving service quality: learning from manufacturing', *International Journal of Production Economics*, 30 (31), July, 1993
MINER *The Management Process Theory, Research and Practice*, Collier Macmillan, 1978
MINTZBERG 'The manager's job: folklore or fact', *Harvard Business Review*, July–Aug., 1975
MINTZBERG *The Nature of Managerial Work*, Prentice Hall, 1980
MINTZBERG *Power in and Around Organisations*, Prentice Hall, 1983
MONCZKA and MORGAN 'Benchmarking: what you need to do to make it work', *Purchasing*, 114 (1), January, 1993
MONTGOMERY *Statistical Quality Control*, Wiley, 1985
MORGAN *Images of Organisations*, Sage, 1986
MUKHI, HAMPTON and BARNWELL *Australian Management*, McGraw-Hill, 1988
NEL and PITT 'Service quality in a retail environment: closing the gaps', *Journal of General Management*, 18 (3), Spring, 1993
NULAND 'Prerequisites to implementation', *Quality Progress*, June, 1990
OAKLAND *Total Quality Management*, Butterworth Heinemann, 1989
O'DELL 'Sharing the productivity payoff', in Werther et al. (eds), *Productivity Through People*, West Publishing, 1986
O'HARA and FRODEY 'A service quality model for manufacturing', *Management Decision*, 31 (8), 1993
PARKA 'Are you tough enough to really lead quality?', *Tapping the Network Journal*, 4 (2), Summer–Winter, 1993
PFEFFER *Power in Organisations*, Pitman, 1981
PINDER *Work Motivation: Theory, Issues, and Applications*, Scott, Foresman, Glenview, 1984
PIRSIG *Zen and the Art of Motorcycle Maintenance*, Bantam, 1974
PRYOR and KATZ 'How benchmarking goes wrong (and how to do it right)', *Planning Review*, 21 (1), Jan./Feb., 1993
RADFORD *The Control of Quality in Manufacturing*, Ronald Press, 1922
REILEY 'The circular organisation: how leadership can optimise organisational effectiveness', *National Productivity Review*, 13 (1), Winter, 1993/1994
RICE 'The Hawthorne defect: persistence of a flawed theory', *Psychology Today*, Feb., 70–74, 1982
ROBBINS and MUKERIJ *Managing Organisations: New Challenges and Perspectives*, Prentice Hall, 1990
ROBERTSON *Quality Control and Reliability*, Pitman, 1971
ROCHE *National Survey of Quality Control in Manufacturing Industries*, National Board of Science and Technology, Ireland, 1981
SABEL *Work and Politics*, Cambridge University Press, 1982
SAUNDERS and WALKER 'TQM in tertiary education', *International Journal of Quality and Reliability Management*, 8 (5), 1991
SAWADA and CALLEY 'Dissipative structures — new metaphors for becoming in education', *Educational Researcher*, 14 (3), 1985
SCHEIN *Organisational Culture and Leadership*, Jossey Bass, 1985

SCHEIN *Process Consultation*, Vol. 2, Addison-Wesley, 1987
SHAW *Group Dynamics: The Psychology of Small Group Behaviour*, McGraw-Hill, 1981
SHEEHY 'Quality leadership: what does it look like?', *National Productivity Review*, 13 (1), Winter 1993/1994
SHETTY 'Aiming high: competitive benchmarking for superior performance', *Long Range Planning*, Feb., 1993
SHEWHART *Economic Control of Quality of Manufactured Product*, Van Nostrand, 1931
SIMS and DEAN 'Beyond quality circles: self-managing teams', *Personnel*, Jan., 1985
SKINNER *Contingencies of Reinforcement: A Theoretical Analysis*, Appleton–Century–Crofts, 1969
SNEE 'Creating robust work processes', *Quality Progress*, February, 1993
SOROHAN 'Open learners', *Training and Development*, 48 (5), May, 1994
SPENDOLINI 'How to build a benchmarking team', *Journal of Business Strategy*, March/April, 1993
STAUB 'A culture of leadership', *Executive Excellence*, 10 (2), Feb., 1993
STEERS and PORTER *Motivation and Work Behaviour*, McGraw-Hill, 1987
STEWART *Contrast in Management*, McGraw-Hill, 1976
STONER and FREEMAN *Management*, Prentice Hall, 1989
SZILAGYI and WALLACE *Organisational Behaviour and Performance*, Goodyear, 1980
TAGUCHI and CLAUSING 'Robust quality', *Harvard Business Review*, Jan.–Feb., 1990
TANNENBAUM and SCHMIDT 'How to choose a leadership pattern', *Harvard Business Review*, 51, May–June, 1973
TANNENBAUM, WESCHLER and MASSARIK *Leadership and Organisation*, McGraw-Hill, 1961
TAYLOR *The Principles of Scientific Management*, Harper and Bros, 1915
TERRY and FRANKLIN *Principles of Management*, Irwin, 1982
THOMAS *Conflict and Conflict Management*, 1976
THOMAS *Total Quality Training*, McGraw-Hill, 1992
TILLES 'The manager's job: a systems approach', *Harvard Business Review*, 41 (1), 73–81, Jan.–Feb., 1963
TOMPKINS 'Team-based continuous improvement: how to make the pace of change work for you and your company (Part 1)', *Materials Handling Engineering*, 48 (1), Jan., 1993
TSIAKALS 'Management team seeks quality improvement from quality costs', *Quality Progress*, 16 (4), 1983
TUCHMAN 'The decline of quality', *New York Times Magazine*, Nov. 2, 1980
TUCKMAN 'Developmental sequence in small groups', *Psychological Bulletin*, 63, 1965
VAZIRI 'Questions to answer before benchmarking', *Planning Review*, 21 (1), Jan./Feb., 1993
VROOM and YAGO *The New Leadership: Managing Participation in Organisations*, Prentice Hall, 1988
WAHBA and BRIDWELL 'Maslow reconsidered: a review of research on the need hierarchy theory', *Organisational Behaviour and Human Performance*, 16, 1976
WALLACE, *CIO*, Vol. 6, No. 12, 1993
WALTON *The Deming Management Method*, Perigee, 1986
WALTON and DUTTON 'The management of interdepartmental conflict: a model and review', *Administrative Science Quarterly*, March, 1969
WATSON 'Resistance to change', in *Concepts for Social Change*, Co-operative Project for Educational Development Series, Vol. 1, National Training Laboratories, 1966
WEATHERLY 'Staff cuts aren't necessarily the best competitive option: benchmarking for better productivity', *Business Mexico*, April, 1993
WEINSHALL *Culture and Management: Selected Readings*, Penguin, 1977

WIDMAN 'Techniques that momentum graphics uses to balance the people and technical sides of quality', *National Productivity Review*, 13 (1), 1993/1994

WILLIAMS and ZIGLI 'Ambiguity impedes quality in the service industry', *Quality Progress*, July, 1987

WOODCOCK and FRANCIS *Team Development Manual*, Gower, 1979

WOODWARD *Industrial Organisation*, Oxford University Press, 1965

WREN *The Evolution of Management Thought*, Wiley, 1979

YOUNG 'Management: checking performance with competitive benchmarking', *Professional Engineering*, 6 (2), Feb., 1993

YUKL *Leadership in Organisations*, Prentice Hall, 1989

Name Index

Abed 278
Adair 173, 180
Albrecht 117, 118, 120, 279
Alderfer 157–8
Atkinson 45, 46, 50

Babbage 19
Barkdull 138
Barrow 148
Bartol 98, 155
Bass 145, 152, 170
Beer 212
Bell 166, 176, 208
Bemowski 104, 107
Benne 208
Berger 205
Berne 166
Bion 220
Blake 147, 150, 215
Blanchard 149
Bogan 108
Bowen 118
Brelin 108
Bridwell 157
Brown 30, 204, 209, 210, 215, 220
Buchanan 176
Buckingham 305
Burns 19, 34, 135, 152, 222
Bush 205, 209, 224

Caldwell 198
Calley 306
Camp 107, 108
Chang 97
Chapman 303
Chase 118
Child 134, 138
Chin 208
Churchman 32
Clausing 122, 126, 129
Cook 55
Crosby 62, 70–3, 82, 86, 114
Crow 107
Curtis 57

Dakin 194
Dale 47, 55, 277, 278, 280
Dean 200
Deming 47, 62, 64–70, 74, 97, 114, 154, 249, 273, 277

Denburg 118
Dessler 139, 151
Drucker 114, 143
Drury 233
Duncalfe 280
Durgesh 104
Dutton 178

English 108
Eureka 122, 126
Evan 178
Evans 104

Fayol 23
Feigenbaum 62, 75, 143
Fiedler 151, 152
Follett 28
Ford 21, 106, 135
Francis 176, 177
Fraser-Robinson 55, 115, 122, 124
Freeman 143
French 144, 166, 176, 208, 225, 226
Friedman 21
Frodey 117
Fullan 205, 207, 212

Gantt 22
Garvin 40, 44, 62, 70, 80, 81, 82, 86, 87, 122, 126, 278
Gilbreth's 22
Greiner 207, 219
Griffin 11
Gryna 54, 96, 101, 232, 233, 249, 269, 280
Guest 12

Hackman 137, 172
Hampton 117
Harvey 204, 209, 210, 215, 220
Haserot 108
Hauser 122, 126, 129
Hequet 109
Hersey 149
Hertzberg 137, 158, 223
Hickman 17
Hofstede 220
Hounshell 42
House 150
Huczynski 176

Irvine 180

NAME INDEX

Ishikawa 48, 62, 73–5, 247
Ivancevich 95

James 26
Janis 174, 178
Johnson 220, 221
Juran 54, 62–4, 85, 96, 100, 101, 114, 122, 232, 233, 249, 269, 280

Kanter 204, 205, 225
Kast 32, 223
Katz 13, 104, 119
Keen 303
Kleiner 118
Koontz 33, 180
Kotler 115, 116, 121
Kotter 12

Labovitz 97
Levacic 233
Lewin 212, 213
Likert 167
Lilley 291
Locke 161
Long 144
Louis 207
Lovelock 118
Luckman 205
Luthans 33

Main 104
Makin 194
Manz 154
Marshall 294
Martin 98, 155
Maslow 31, 156–7
Mayo 29, 30
McCarthy 121
McClelland 158–60
McGregor 31, 32, 177
Mefford 117
Miles 207
Miner 178
Mintzberg 11, 15, 16, 144, 173, 183
Monczka 106
Montgomery 249, 250, 260, 262, 276
Morgan 19, 32, 63, 106, 178
Mouton 145, 150, 215
Mukerij 11
Mukhi 139, 143
Münsterberg 28, 29

Nel 119
Nuland 99

O'Dell 189
O'Hara 117
Oakland 45, 49, 51, 63, 89, 305

Oldham 137
Owen 18

Parka 154
Pirsig 81
Pfeffer 144
Pinder 160
Pitt 119
Plunkett 47, 55, 277, 278
Porter 156, 158, 159
Pryor 104

Radford 42
Randolph 135
Raven 144, 225, 226
Reiley 143
Rice 30
Robbins 11
Robertson 194, 278
Roche 280
Roethlisberger 30
Rooney 305
Rosenzweig 32, 223
Ryan 122, 126

Saunders 58, 306
Sawada 306
Schein 216, 220
Schmidt 149
Schneider 118
Scholes 220, 221
Shaw 173
Sheehy 143
Shetty 104, 105, 107
Shewhart 249
Silva 17
Sims 154, 200
Skinner 220
Smith 19
Snee 119
Sorohan 144
Spendolini 107
Stalker 19, 34, 135, 222
Staub 144
Steers 158, 159
Stewart 12
Stoner 143
Szilagyi 176

Taguchi 62, 75–6, 86, 122
Tannenbaum 143, 149
Tannock 304, 305
Tansik 118
Taylor 19, 20, 136
Thomas 55, 178
Tilles 33
Tompkins 143
Tsiakals 281

Tuchman 81
Tuckman 169, 305

Van Epps 107
Vaziri 105, 108
Vroom 152–3

Wahba 157
Walker 58, 306
Wallace 176, 233
Walton 65, 178
Watson 210
Weatherly 108

Weinshall 220
Widman 196
Williams 117
Woodcock 176, 177
Woodward 34

Yago 152–3
Young 107
Yukl 143

Zemke 117, 118, 120, 279
Zigli 117

Subject index

Acceptance plans 265
Acceptance sampling 264
Accreditation 295
Auditing to BS EN ISO 9000 299

Benchmarking
  benefits 108
  definition 104
  and goal-setting 105
  influences 107
  limitations 109
  performance indicators 106
  process 107
  types 106
BS 4778 276, 282
BS 6143 276, 282
BS 7750 302–3
BS 7850 301
BS EN ISO 9000 45, 59, 86, 125, 237, 276, 281, 282
  benefits of certification to 294
  origin and basis 285
  parts of 287

Centralization versus decentralization 139
Certification and accreditation 291
  accreditation 295
  certification 292
  definitions 292
  process 298
Change
  agent 214
  causes 205
  definitions 204
  dealing with 206
  implementation 208, 212
  interventions 218
  managing 206
  nature 204
  and organizational life-cycle 219
  process consultation 216
  resistance 209
Communication
  group configurations 184
  importance 180
  process 183
  types 182
Conflict 177
  managing conflict 179

Conformance tools 122
Control
  definition 230
  process requirements 232
  systems 230
Crosby's
  five absolutes of quality 70
  fourteen point plan 71
Culture
  definition 219
  change
    and power 221
    and politics 223
Customer 115
Customer-supply model 83, 124

Deming's
  cycle 47
  deadly diseases 68
  fourteen points 65
Design methods
  quality function deployment 126

Effects of
  differing quality views 83
  standards on traceability 125
Eight dimensions of quality 84
European quality award 304

Factors affecting customer perception 84
Feigenbaum's hidden plant 75
Five systems of TQM 48
Four eras of quality management 40

Garvin's five quality bases 81
Group
  characteristics 166
  development 169
  effectiveness and efficiency 172
  norms and conformity 174
  process 173
  types 167
    formal 167
    informal 168

Human resource management
  compensation 197
  definition 189
  and TQM 189

SUBJECT INDEX

Human resource management (*continued*)
  planning 191
  recruitment 192
  selection 193
  training and education 195
  performance appraisal 197
  workforce relationships 198

Inspection
  process 247
    measurement 247
    quality measurements 248
    variation 248
Investors in people (IIP) 303
Integrated total quality management 315

JIT methods 236
Job design
  methods 136
Juran's
  five quality characteristics 62
  four elements of fitness for use 63

Leadership
  definition 143
  influence 144
  theories
    trait qualities of leadership 145
    behavioural qualities of leadership 146
    situational qualities of leadership 148
    self-leadership 153

Management
  definition 10
  levels 11
  functions 24
  process 26
  theories
    behavioural 27
    classical 19
    contingency 33
    human relations 31
    pre-behavioural 28
    systems 32
Managerial
  skills 13
  roles 14
Marketing
  customer-oriented 123
  definition 115
  and design 114
  mix 121
  planning 121
Materials control and prevention 236
Motivation
  nature 155
  theories 155
    need 156

    cognitive 160
    reinforcement 162

Operating characteristic curve 268
  construction 269
Organizing 132
Organizational
  design 134
  effectiveness—implications 140
  structure 133

Process
  capability uses 269
  variability reduction 271
  of implementation of TQM 52

Quality
  action plans 102
  circles 199
  environmental analysis 98
  goals 100
  mission 98
  planning
    need 95
    responsibility 95
    process 97
  plans
    types 96
  policy 99
  performance evaluation 103
  related costs
    classification 276
    hidden costs 279
    importance 277
    life cycle 280
    management of 280
    why measure them? 278
  service
    benefits 120
    culture 118
    effectiveness 118
    management 119
  strategy implementation 102
  system 284
Quality control charts
  variable control charts 253
  attribute control charts 256
  cumulative summation charts 260
  Shewhart and CuSum charts 263
Quality development through
  inspection 40
  quality assurance 44
  quality control 43
  total quality management 45
Quality function deployment 126
Quality loss function (Taguchi) 272

Quality management
  writers
    Crosby 70
    Deming 64
    Feigenbaum 75
    Garvin 70
    Ishikawa 73
    Juran 62
    Taguchi 75
  Standards
    BS EN ISO 9000 284, (elements) 310, (survey instrument) 311
    BS 7750 302
    BS 7850 301
    EN 45012 313
    European Quality Award 304
    Investors in People 303
    ISO 8042 286
Questions
  what is a group? 166
  what is quality planning? 96
  who is responsible for quality planning? 95
  who is the customer? 115
  why plan? 94
  why the need for planning for quality? 95

Sampling plans
  single sampling plan 267
  double sampling plan 268

Seven old tools 237–42
Seven new tools 237, 242–3
Six steps of problem solving 64
Statistical process control
  definition 246
  and quality improvement 270
Statistical control charts
  application 250
  concept 249
  definition 249
  steps in development 252
  types 253
Strategic quality
  goals 100
  leadership — need 144

Teambuilding 176
Three quality views 80
Total quality management
  definition 45
  implementation issues 48, 51
  practices — the reality 55
  process of implementation 52
  sector experiences
    manufacturing 55
    services 57
Traceability 123
  effects of standards on 125
Types of quality plans 96